Engineering of Power Plant and Industrial Cooling Water Systems

Engineering of Power Plant and Industrial Cooling Water Systems

Charles F. Bowman

Seth N. Bowman

CRC Press
Taylor & Francis Group
Boca Raton London New York

CRC Press is an imprint of the
Taylor & Francis Group, an **informa** business

First edition published 2022
by CRC Press
6000 Broken Sound Parkway NW, Suite 300, Boca Raton, FL 33487-2742

and by CRC Press
2 Park Square, Milton Park, Abingdon, Oxon, OX14 4RN

CRC Press is an imprint of Taylor & Francis Group, LLC

ISBN: 978-0-367-77528-5 (hbk)
ISBN: 978-1-032-00039-8 (pbk)
ISBN: 978-1-003-17243-7 (ebk)

Typeset in Times
by Straive, India

To Nancy
Without whose help this book could not have been possible

Contents

Preface

Although this book draws heavily on lessons learned from the authors' experience in the nuclear power industry, they are equally applicable to many other types of power plants as well as to industrial plants. From 1945 to 1975, when most of the existing electric power plants were designed and constructed, engineering of the plant cooling water system was considered a low priority item. These systems made little or no contribution to the actual production of electricity and their failures rarely resulted in a plant outage. Established engineering handbooks such as the *Cameron Hydraulic Data Book* and the *Hydraulic Institute Standards* were employed in a cook book fashion to arrive at more-or-less standard designs. All of that started to change as the current fleet of nuclear power plants began operating in the 1960s and 70s.

The Tennessee Valley Authority (TVA) entered the nuclear power industry in 1966 by beginning construction of Unit 1 of the Browns Ferry Nuclear Plant, an essentially turnkey 1,100 MWe boiling water reactor plant designed by General Electric. The TVA fleet was expanded with the start of construction of Unit 1 at Sequoyah (1970), followed by Watts Bar Unit 1 (1973). In September 1973, the TVA Division of Engineering Design was reorganized with design projects dedicated to each nuclear project and engineering branches to support the projects. As part of that reorganization, Charles F. (Chuck) Bowman, a graduate of the University of Tennessee (BSME in 1964 and MSME in 1966), was selected to supervise a section in the Mechanical Engineering Branch that was responsible for engineering the cooling water systems for all of the TVA nuclear plants. Chuck is indebted to the outstanding engineers who worked for him during that period and developed much of the technology described herein. In 1986, he was promoted to be the corporate specialist for thermal performance and cooling water systems, supporting all seven of the TVA operating nuclear units.

In 1976, when Chuck reviewed the results of preoperational flow tests conducted on Browns Ferry Unit 1, he realized that these systems had major problems. The tests showed that critical heat exchangers were not receiving the required cooling water flow. What followed was years of learning lessons of how to deal with issues such as microbiologically induced corrosion and Asiatic clams. Based on his experience and that of others, a better way of engineering these systems for both nuclear and fossil power plants as well as industrial plants is possible in the future.

In 1994, Chuck retired from the TVA to form Chuck Bowman Associates, Inc., serving the electric power industry and specializing in cooling water systems and the analysis and testing of pumping systems, heat exchangers, cooling towers, spray ponds, etc., as well as the thermal aspects of turbine cycle systems. The purpose of this book is to impart to the next generation knowledge and experience received during more than 50 years of working in the field of cooling water systems.

The term cooling water systems (sometimes referred to as service water systems) refers principally to the systems required to cool the many heat exchangers used in all power and industrial plants. Although many of the lessons learned also apply to the large condenser circulating water system that condenses the steam coming from the main condenser, that system is not the primary emphasis of this book.

Authors

Charles F. (Chuck) Bowman, P.E., is the president of Chuck Bowman Associates, Inc. (CBA), an engineering consulting firm serving the electric power industry since 1994. Chuck received his BS and MS degrees in Mechanical Engineering from the University of Tennessee, and he is a registered professional engineer in Tennessee. CBA specializes in engineering analysis of electric power generating cycles and related fields including the design and analysis of heat exchangers, cooling towers, spray ponds, cooling water systems, etc. Before forming CBA, Chuck was with the Tennessee Valley Authority (TVA) for 28 years where he supervised the engineering of the cooling water systems for all of the TVA nuclear plants. Prior to his retirement from TVA, he was the senior engineering specialist for thermal performance and cooling water systems in TVA's Corporate Engineering Office. Chuck has served as a consultant to the Electric Power Research Institute (EPRI). He is the author of EPRI Report No. 1007248, *Alternative to Thermal Performance Testing and/or Tube-Side Inspections of Air-to-Water Heat Exchangers*; EPRI Software Manual No. 3002005344, *Turbine Cycle Equipment Evaluation Workbook,* and was a paid reviewer of EPRI Report No. 1021065, *Heat Exchanger Performance Analysis.* Chuck serves on the American Nuclear Society (ANS) Working Group ANS 2.21, Criteria for Assessing Atmospheric Effects on the Ultimate Heat Sink, and has served on American Society of Mechanical Engineers (ASME) committees that authored ASME PTC 23.1 – 1983, *Code on Spray Cooling Systems* and ASME PTC 12.5 - 2000, *Single Phase Heat Exchangers* and is a contributing editor to *Marks' Handbook*, 12th Edition. His most recent publication is *Thermal Engineering of Nuclear Power Stations: Balance-of Plant Systems*, published by CRC Press.

Seth N. Bowman received his BS and MS degrees in Chemical Engineering from the University of Tennessee. He currently serves as the senior manager of the corrective action program at the Y-12 National Security Complex, managed and operated by Consolidated Nuclear Security, LLC. In prior roles, Seth has been responsible for the assessment function for the Engineering Division at Y-12 and also served as a shift technical adviser and shift manager for Building 9212. He is co-author of *Thermal Engineering of Nuclear Power Stations: Balance-of Plant Systems*, published by CRC Press.

List of Abbreviations and Acronyms

ACRS	Advisory Committee on Reactor Safeguards
ASME	American Society of Mechanical Engineers
AWHX	air-to-water heat exchanger
AWWA	American Water Works Association
BCDMH	bromochlordimethylhydantoin
BFNP	Browns Ferry Nuclear Plant
BTA	best technology available
C	Colbert
CCW	condenser circulating water
CFU	colony-forming units
CGS	Columbia Generating Station
CML	cement-mortar lining
CP&L	Carolina Power & Light
CPP	concrete pressure pipe
CS	carbon steel
CU	Cumberland
CWA	Clean Water Act
EECW	emergency equipment cooling water
EPA	Environmental Protection Agency
ERCW	essential raw cooling water
FBSP	flat bed spray pond
FRP	fiberglass-reinforced plastic
G	Gallatin
HDPE	high density polyethylene
HX	heat exchanger
HXs	heat exchangers
IAM	Institute for Applied Microbiology
IN	Information Notice
INPO	Institute of Nuclear Power Operations
IPS	intake pumping station
JS	John Sevier
K	Kingston
MC	main condenser
MIC	microbiologically induced corrosion
MPY	mils per year
NDCT	natural draft cooling tower
NDE	nondestructive examination
NPDES	National Pollution Discharge Elimination System
NRC	Nuclear Regulatory Commission
OSCS	oriented spray cooling system

PDI	pre-lined ductile iron
PHX	plate heat exchanger
PVC	polyvinyl chloride
QA	quality assurance
RCW	raw cooling water
RT	radiographic testing
SER	Safety Evaluation Report
SI	standard international
SM&E	Singleton Materials & Engineering Laboratory
SNP	Sequoyah Nuclear Plant
SRB	sulfur-reducing bacteria
SS	stainless steel
STHX	shell-and-tube heat exchanger
SWAP	Service Water Assistance Program
TEMA	Tubular Exchange Manufacturer's Association
TRC	total residual chloride
TVA	Tennessee Valley Authority
UT	ultrasonic testing
WB	Watts Bar
WBNP	Watts Bar Nuclear Plant
WBT	wet-bulb temperature
WC	Widows Creek

1 Introduction

1.1 BACKGROUND

The overall operating and safety record of the existing nuclear industry is outstanding, but as one will see in this introductory chapter, the record of the engineering of the cooling water systems has been less than sterling, as indicated by the public records maintained by the Nuclear Regulatory Commission (NRC). The purpose of this chapter is not to point fingers or rehash old problems but to learn from them so that the engineering of future industrial and electric power plants, both fossil and nuclear, will not repeat the same mistakes.

1.2 TYPES OF COOLING WATER SYSTEMS

The main condenser circulating water (CCW) system is not technically a "cooling" water system, since that system serves to condense saturated steam exiting from the low pressure turbine and that condensation process is isothermal, although the CCW is heated in the process. For the vast majority of electric power plants, the CCW system is either an open once-through or open recirculating system, in which case the waste heat from the main condenser (MC) is rejected to the environment via a cooling tower, principally by evaporation, and the lost water must be replaced by some "makeup" source. In some power plants, the cooling water system operates in parallel to the MC, drawing its cooling water from the CCW system and returning it thereto. In other plants, the cooling water may be drawn directly from some river, lake, or ocean and passes straight through the plant heat exchangers (HXs) and is discharged back to its source or may serve as makeup to a recirculating CCW system. In some plants, the cooling water system may be a closed recirculating system, rejecting its heat to the environment via cooling towers, cooling pond, or spray pond.

1.3 TYPES OF COOLING WATER SYSTEM PUMPS

Cooling water pumps are either of the vertical wet pit mixed flow type with a motor mounted above the pump column and a line shaft extending down to the pump bowl(s) submerged in the water source or they are horizontal centrifugal pumps located in a dry well. Horizontal centrifugal pumps are normally located inside a building, taking suction from the CCW system or as a booster pump in the cooling water system. Vertical mixed flow type pumps are normally located in an intake pumping station (IPS), the structure located on the river, lake, or ocean that is the source of cooling water. (See Figure 3.1.) Most CCW pumps are of the vertical turbine type variety, but in some power plants with cooling towers, the CCW pumps may be horizontal centrifugal pumps located in a dry well near and below the cooling tower basin.

Cooling water pump impeller, line shaft, and coupling failures are a continuing problem area. In 1994 the NRC issued Information Notice (IN) 94-45[1] to inform

1

licensees of a problem of high vibration levels on one of the cooling water pumps at the Grand Gulf Nuclear Station. Disassembly of the pump showed extensive corrosion of the carbon steel (CS) bolts and lock washers used in the pump shaft coupling assembly. A survey of cooling water pump failures conducted by *Operating Experience Digest* between 1998 and 2005 reported more than one failure per year, with several nuclear plants experiencing more than one failure. The most frequent cause was reported to be corrosion of shaft and/or bolting material. Pump performance degradation occurs due to normal wear, requiring clearance adjustments or overhaul of the pump impeller/bowl assembly.

1.4   INTAKE PUMPING STATIONS

Although the design of an IPS is a function of the geography of the power plant site, almost all designs incorporate devices to restrict the inflow of unwanted material. The first barrier is normally a trash rack with a rake for periodically removing unwanted materials. Beyond the trash rack might be a stop log to permit the unwatering of the sump. The next barrier would be a traveling water screen with a screen wash pump that takes suction downstream of the traveling screen. Finally, there may be an assortment of pumps including the CCW pumps, the cooling water pumps, high pressure fire protection pumps, cooling tower makeup pumps, and miscellaneous service water pumps. The sump for each pump should be designed to promote smooth flow of water to the individual pump suction. Normally, the motors driving these vertical wet pit pumps are located on the deck above.

Overwhelming of the features engineered to permit the free flow of water to the pumps and protect them from the intrusion of foreign objects such as debris, fish and other aquatic animals, algae, marsh grass, seaweed, leaves, ice, and cooling tower fill material has been a major impediment to safe and reliable operation of service water systems. In 1989 the NRC initiated Generic Letter 89-13,[2] a review of cooling water system events/failures that indicated that approximately 10 percent of these were due to foreign material intrusion. In 1990, NRC issued IN 92-49[3] that reported on debris accumulation on the traveling water screens at the following nuclear stations:

- Fitzpatrick Nuclear Plant, debris caused shear pins on both screens to fail and causing the screen to bow inward and allowing some of the debris to pass around the screens
- Millstone Nuclear Plant Unit 1, debris caused three of the five screens to collapse
- Arkansas Nuclear One, debris caused the screens to fail causing the screen to bow inward and allowing the debris to pass around the screens, clogged strainers on the discharge of the cooling water pumps on both operating loops.

More recently, nuclear plants have reported a large ingress of marsh grass at Hope Creek and Oyster Creek Generating Stations, a massive algae problem at Pickering Nuclear Plant, and frazil ice blocking the intakes of other plants. (Frazil ice is a collection of loose, randomly oriented ice crystals formed in supercooled turbulent water.) Clearly, the current design of intake structures makes them vulnerable to site-specific intake blockage scenarios.

1.5 COOLING WATER SYSTEM PIPING AND VALVES

With the exception of CCW piping, the vast majority of cooling water system piping is CS. CCW piping, which is huge by comparison with cooling water system piping, is typically concrete pressure pipe (CPP) and/or square cast-in-place concrete except above ground within the IPS and turbine building where CS is used. Due to dismal experience with using CS piping in cooling water systems, CS piping is being replaced piecemeal with materials such as stainless steel (SS), copper, polyvinyl chloride (PVC), fiberglass-reinforced plastic (FRP), and high-density polyethylene (HDPE). In situ cement mortar lining of large buried CS piping has proven to be effective but is not widely in use. In some instances, epoxy coatings have been applied with mixed results. The surface must be prepared to bright metal, and any imperfection results in preferential galvanic pitting corrosion attack. CS valves, being thicker and of more robust material, generally provide better service in cooling water systems than CS piping when unobstructed.

The problem of cooling water system piping leaking was recognized by the NRC with issuance of Bulletin 80-24[4] as early as 1980 when significant multiple piping leaks at the Indian Point Nuclear Power Station Unit 2 resulted in the discovery of several inches of water on the floor inside the containment of the reactor. In 1985, the NRC issued IN 85-30[5] to report that pinhole leaks were discovered in SS cooling water piping at Unit 2 of the H. B. Robinson Nuclear Plant. In 1994, leaks were reported in the cooling water piping serving the emergency diesel generators at the Connecticut Yankee and Beaver Valley Nuclear Power Plants. In IN 94-79,[6] the NRC reported that these leaks were due to microbiologically induced corrosion (MIC). In 2006, a leak was detected in a portion of the cooling water system piping buried at Oyster Creek Generating Station. As a result of these and other cooling water system piping failures, the NRC issued Generic Letter 90-05[7] as guidance for performing temporary repairs of the piping.

In 1983, the NRC reported in IE IN 83-46[8] that seven of eight motor-operated butterfly valves in the cooling water system at the Surry Power Station failed to open during a test. The failures were attributed to marine growth, among other reasons. In 1994, The NRC reported that 15 of 31 cooling water system check valves were found to be potentially inoperable due to the accumulation of silt and corrosion products at the Fitzpatrick Nuclear Plant.[9] In 1991, the NRC issued report IN 94-61[10] stating that the stem assembly on a gate valve in the cooling water system at the Farley Nuclear Plant failed due to corrosion.

In 1985, the NRC issued IN 85-24[11] to report the discovery of delamination and peeling of the epoxy lining of the Palo Verde Nuclear Generating Station's cooling water system. In 1997, the NRC issued Licensee Event Report 97-037[12] on pieces of the PVC liner material found in the strainer basket serving the emergency diesel generator HXs at the Millstone Nuclear Power Station, Unit 2.

1.6 COOLING WATER SYSTEM FLOW DEFICIENCIES

In 1981, the NRC issued Bulletin 81-03[13] to report that the Resident Inspector discovered that there was inadequate cooling water flow to the reactor containment cooling units at Unit 1 of the Arkansas Nuclear One, due to plugging of the heat exchanger (HX) by Asiatic clams (Corbicula). Also in 1981, the NRC issued IN 81-21[14] to report on a growth of some form of sea mollusk that was discovered on the cooling water piping serving the component cooling HX in Unit 1 at the San Onofre Nuclear Generating Station. The mollusks impaired movement of the butterfly valves used to isolate the HX, reduced the cross-sectional diameter of the cooling water piping, and blocked the HX tubes. As a result, the cooling water flow rate to the HX was reduced. In that same year, the NRC issued IN 81-21[14] to report that the baffle plate in the U-tube type residual heat removal HX at Unit 1 of the Brunswick Nuclear Plant was damaged, allowing the cooling water to bypass the tube bundle. A flow test indicated that the HX was unable to remove the decay heat because oyster shells had blocked the inlet side of the HX tubes. Similar events occurred with the reactor building closed cooling water HX at the Pilgrim Nuclear Power Station. In 1986, the NRC issued IN 86-96[15] to report that the HXs cooling the gearbox on the charging pumps that are part of the emergency reactor core cooling system at Unit 1 of the Farley Nuclear Plant failed to adequately cool the gearboxes, causing them to overheat due to silt in the cooling water system. A similar event occurred at Unit 2 of the Salem Nuclear Power Plant.

In 1988, the NRC issued IN 88-37[16] to report that when Unit 2 of the Catawba Nuclear Plant was at 20% power after the first refueling outage, the feedwater regulating valve failed open, causing a main turbine to trip off line due to a high-high level in one of the steam generators. When both auxiliary feedwater pumps started automatically, low suction pressure resulted in an automatic swap over to the cooling water system, resulting in raw cooling water (RCW) from Lake Wylie being pumped into the steam generators. When the water level in the steam generators dropped to their low-low level set point, the reactor tripped automatically. After normal feedwater flow was established, the auxiliary feedwater control valves were found to be clogged with Asiatic clams. A similar event occurred at the Davis-Besse Nuclear Power Station in 1985. Catawba has also experienced serious corrosion and debris buildup in their cooling water system, challenging the ability of the system to provide the required flow during an accident.

In 1989 the NRC issued IN 89-76[17] concerning a presentation at an EPRI conference by representatives of the Ontario Ministry of the Environment, Detroit Edison Company, and the U.S. Fish and Wildlife Service on zebra mussels. Bio-fouling due to zebra mussels was observed at several power plants, water treatment plants, and industrial facilities along Lake Erie where they threatened operation of the condensers and cooling water and fire protection systems.

Also in 1989, the NRC issued IN 90-39[18] to report on inadequate cooling water flow to safety-related HXs at several nuclear stations. At the Farley Nuclear Plant, flow was inadequate to certain safety-related HXs and required operator action. At the Peach Bottom Power Plant, Unit 2, flow was found to be inadequate due to the accumulation of silt and corrosion products in the cooling water piping. At Fitzpatrick Power Plant, silt was found in check valves in the cooling water system lines to the seal water coolers for residual heat removal pumps that could have prevented the system from performing its safety function.

1.7 HEAT EXCHANGER FOULING

In 1986, the McGuire Nuclear Station reported to the NRC that over the years a combination of organic and inorganic compounds have fouled a number of HXs, including the containment spray, component cooling water, and control room chillers.[15] In 1996 in Licensing Event Report 96-001-01[19] the staff at Unit 1 of the Calvert Cliffs Nuclear Power Plant reported to the NRC that they had determined that their safety-related HXs served by the cooling water system may not have been capable of meeting their intended safety function during periods of high Chesapeake Bay water temperatures, due to microfouling on the tube side of the HX. In 2007, the NRC issued IN 2007-28[20] to report that when the Palo Verde Nuclear Generating Station noted elevated temperatures in the intake air for one of the emergency diesel generators, a lotion-like substance was found to be fouling the cooling water side of the diesel generator HXs.

1.8 ENGINEERING DESIGN ERRORS

In 1990, the NRC issued IN 90-26[21] to report that after conducting extensive flow and pressure drop tests on the cooling water side of room coolers at the Clinton Power Station, the staff found that the relationship between the flow and pressure drop provided by the manufacturer for these room coolers was wrong and that the actual pressure drop was greater than advertised. The same manufacturer had supplied room coolers to approximately 50 nuclear power plants.

A survey of cooling water pump failures conducted by *Operating Experience Digest* concluded that the pump failures were due to the specification of incorrect bolting and shaft material, improper heat treatment, and the use of dissimilar metals resulting in galvanic corrosion. Some pumps were designed with the resonant frequency of the line shaft being too close to the operating speed of the pump.

In 1996, the NRC issued Generic Letter 96-06[22] when the staff at the Diablo Canyon Power Plant determined that the component cooling water serving the reactor containment air coolers located high inside the reactor building's containment could flash to steam in the cooling coils during an accident. The discovery of this engineering design error caused the Westinghouse Electric Corporation to issue a nuclear safety advisory letter. The same problem was identified at the Connecticut Yankee Nuclear Power Plant where the containment air coolers are served by the cooling water system. In that same year, a similar condition was discovered at the Point Beach Nuclear Plant and at Salem Nuclear Power Plant. When the cooling water flashes to steam, the flow rate is reduced due to the higher pressure drop through the HX and heat removal capability is reduced. In the cases of Diablo Canyon and Salem, cooling water flow ceased completely following the accident for a period of time while pumps were restarted. At Salem, valves were required to operate to isolate non-safety-related HXs. Since the peak containment pressure occurs within seconds following an accident such as a pipe break, the cooling water would not be restored in time to limit the peak containment pressure. Further, an analysis of the cooling water system flows at Connecticut Yankee concluded that restoring the cooling water flow would result in hydrodynamic loads due to water hammer that would exceed the piping and piping support structural limits.

1.9 NRC OVERSIGHT

In 1981, the NRC issued Bulletin 81-03[13] to require nuclear plant licensees to determine whether or not Asiatic clams or other mussels were present in their cooling water systems and if present to confirm that the individual components were receiving adequate cooling water flow. The response to that bulletin indicated that bivalves were present in approximately 45% of the nuclear plant sites in the United States. Serious fouling was reported at several sites as indicated above. Up until that time, some nuclear stations had not ever verified that the HXs were receiving adequate cooling water flow.

In 1988, the NRC published NUREG 1275. Vol. 3,[23] a comprehensive review and evaluation of cooling water system failures and degradations reported in nuclear plants from 1980 to 1987. The report identified 29 events involving significant degradation or failures of the cooling water system, an average of 3.6 events a year.[22]

In 1989, the NRC issued Generic Letter 89-13[2] requiring nuclear plant licensees to do the following:

I. For open-cycle service water systems, implement and maintain an ongoing program of surveillance and control techniques to significantly reduce the incidence of flow blockage problems as a result of bio-fouling.
II. Conduct a test program to verify the heat transfer capability of all safety-related HXs cooled by service water. The total test program was to consist of an initial test program and a periodic retest program of HXs connected to or cooled by one or more open-cycle systems.
III. Ensure by establishing a routine inspection and maintenance program for open-cycle service water system piping and components that corrosion, erosion, protective coating failure, silting, and bio-fouling cannot degrade the performance of the safety-related systems supplied by service water.
IV. Confirm that the service water system would perform its intended function in accordance with the licensing basis for the plant.
V. Confirm that maintenance practices, operating and emergency procedures, and training involving the service water system are adequate to ensure that safety-related equipment cooled by the service water system would function as intended and that operators of this equipment will perform effectively.

It should be noted that Generic Letter 89-13 did not require the nuclear plants to conduct flow balance tests of their cooling water systems. For example, in 1990, when Clinton Power Station measured the actual flows to their room coolers, they found that the flows were from 10% to 80% less than the design flows.[21] Flows to other HXs ranged from 2% to 42% less than the design flows, while other components served by the cooling water system were receiving excess flow up to 213% of the design value.[21]

In 1992, the NRC issued Temporary Instruction 2515/118[24] and initiated a program of conducting operational performance inspections of the cooling water systems for nuclear plants the United States. The inspections included a review of the mechanical systems engineering design and configuration control and nuclear plant operations, maintenance, and testing and quality assurance (QA)/corrective actions. Although the findings of these inspections are too numerous to enumerate here, they did instill in the nuclear industry an awareness of the importance of cooling water systems to nuclear safety. Additional inspections are ongoing.

In 2007, the NRC issued IN 2007-06[25] to report on potential common cause vulnerabilities in essential cooling water systems. Two of the nuclear stations were outside the United States. In the first such case, a manhole pipe broke; in the second a cooling water pipe ruptured, flooding the piping gallery with sea water. In the United States, a 3 gallons per hour through-wall leak in a 30-inch (76.2 cm) pipe was found at Unit 2 of the South Texas Project due to cavitation damage downstream of a heavily throttled butterfly valve, and a through-wall leak was found in the endbell of one of the emergency diesel generator HXs due to MIC.

1.10   AUTHORS' PERSPECTIVE

Over the years, workable solutions to many of the deficiencies in the engineering of cooling water systems have been developed by the authors and others, but in many cases, these solutions have not been widely disseminated. It is the hope of the authors that this book will aid in imparting to the next generation the knowledge and experience gained during more than 50 years of working in the field of cooling water systems.

REFERENCES

1. IN 94-45. *Potential Common-Mode Failure Mechanism for Large Vertical Pumps*, Nuclear Regulatory Commission, Washington, D.C., 1994.
2. Generic Letter 89-13. *Service Water System Problems Affecting Safety-Related Equipment*, Nuclear Regulatory Commission, Washington, DC, 1989.
3. IN 92-49. *Recent Loss or Severe Degradation of Service Water Systems*, Nuclear Regulatory Commission, Washington, DC, 1992.
4. Bulletin 80-24. *Prevention of Damage Due to Water Leakage Inside Containment (October 17, 1980 Indian Point 2 Event)*, Nuclear Regulatory Commission, Washington, DC, 1989.
5. IN 85-30. *Microbiologically Induced Corrosion of Containment Service Water System*, Nuclear Regulatory Commission, Washington, DC, 1985.
6. IN 94-79. *Microbiologically Influenced Corrosion of Emergency Diesel Generator Service Water Piping*, Nuclear Regulatory Commission, Washington, DC, 1994.
7. Generic Letter 90-05. *Guidance for Performing Temporary Non-Code Repair of ASME Code Class 1, 2, and 3 Piping*, Nuclear Regulatory Commission, Washington, DC, 1990.
8. IE IN 83-46. *Common-Mode Valve Failures Degrade Surry's Recirculation Spray Subsystem*, Nuclear Regulatory Commission, Washington, DC, 1983.

9. Third EPRI Balance-of-Plant Heat Exchanger NDE Workshop: A Regulatory Perspective, Davis, J. A., et al, Electric Power Research Institute, Charlotte, NC, 1994.

10. IN 94-61. *Corrosion of William Powell Gate Valve Disc Holders*, Nuclear Regulatory Commission, Washington, DC, 1994.

11. IN 85-24. *Failure of Protective Coatings in Pipes and Heat Exchangers*, Nuclear Regulatory Commission, Washington, DC, 1985.

12. Licensee Event Report 97-037. *Service Water Piping Liner Material Peeled Off and Lodged in the Strainer for the Emergency Diesel Generator Heat Exchangers*, Nuclear Regulatory Commission, Washington, DC, 1997.

13. Bulletin 81-03. *Flow Blockage of Cooling Water to Safety System Components by Corbicula Sp. (Asiatic Clam) and Mytilus Sp. (Mussel)*, Nuclear Regulatory Commission, Washington, DC, 1981.

14. IN 81-21. *Potential Loss of Direct Access to Ultimate Heat Sink*, Nuclear Regulatory Commission, Washington, DC, 1981.

15. IN 86-96. *Heat Exchanger Fouling Can Cause Inadequate Operability of Service Water Systems*, Nuclear Regulatory Commission, Washington, DC, 1986.

16. IN 88-37. *Flow Blockage of Cooling Water to Safety System Components*, Nuclear Regulatory Commission, Washington, DC, 1988.

17. IN 89-76. *Bio-fouling Agent: Zebra Mussel*, Nuclear Regulatory Commission, Washington, DC, 1989.

18. IN 90-39. *Recent Problems with Service Water Systems*, Nuclear Regulatory Commission, Washington, DC, 1990.

19. Licensing Event Report 96-001-01. *SRW Heat Exchanger Micro-fouling Higher Than Assumed in Design Basis*, Nuclear Regulatory Commission, Washington, DC, 1996.

20. IN 2007-28. *Potential Common Cause Vulnerabilities in Essential Service Water Systems due to Inadequate Chemistry Control*, Nuclear Regulatory Commission, Washington, DC, 2007.

21. IN 90-26. *Inadequate Flow of Essential Service Water to Room Coolers and Heat Exchangers for Engineered Safety-Feature Systems*, Nuclear Regulatory Commission, Washington, DC, 1990.

22. Generic Letter 96-06. *Assurance of Equipment Operability and Containment Integrity during Design-Basis Accident Conditions*, Nuclear Regulatory Commission, Washington, DC, 1996.

23. NUREG 1275. Vol. 3. *Operating Experience Feedback Report – Service Water System Failures and Degradations*, Nuclear Regulatory Commission, Washington, DC, 1989.

24. Temporary Instruction 2515/118. *NRC Inspection Manual: Service Water System Operational Performance Inspection*, Nuclear Regulatory Commission, Washington, DC, 1992.

25. IN 2007-06. *Potential Common Cause Vulnerabilities in Essential Service Water Systems*, Nuclear Regulatory Commission, Washington, DC, 2007.

2 Properties of Water

2.1 TEMPERATURE

There are four temperature scales in common use, two in English units and two in standard international (SI) units, The English units are degree Fahrenheit (°F) and degree Rankine (°R). The SI units are degree Celsius (°C) and degree Kelvin (°K). Figure 2.1 shows the temperatures in English units.

Figure 2.2 shows the temperatures in SI units. Virtually all engineering of cooling water systems is performed in either Fahrenheit or Celsius units. The formula for converting from SI units to English units is shown in Equation 2.1

$$°F = 1.8 \times °C + 32 \tag{2.1}$$

2.2 PRESSURE

As seen in Figure 2.3, pressure may be measured in three different ways: (1) gauge pressure (i.e. pressure above atmospheric pressure), (2) vacuum pressure (i.e. pressure below atmospheric pressure), and (3) absolute pressure (i.e. pressure above zero pressure or a perfect vacuum).

Perhaps one of the most common mistakes made in any power plant is confusing gauge pressure with absolute pressure. Most gauges inside a power plant measure absolute pressure. Since the turbine cycle is isolated from the atmosphere, gauge pressure is largely irrelevant. Also, the properties of steam and condensate in a turbine cycle are related to absolute pressure. However, there may be the odd gauge

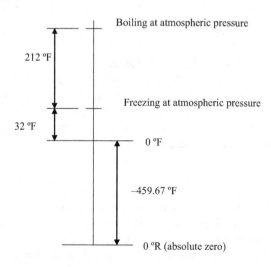

FIGURE 2.1 Temperatures measurements in English units.

9

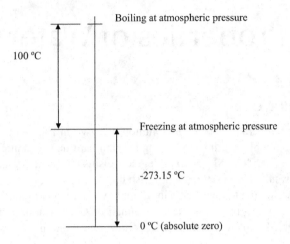

FIGURE 2.2   Temperatures measurements in SI units.

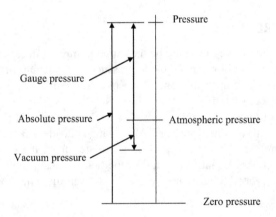

FIGURE 2.3   Pressure measurements.

that reads in gauge pressure. There also may be an occasional instrument that measures vacuum. Be warned! In dealing with cooling systems, almost all instruments read gauge pressure. The units lbf/in^2 (English) and kPa (SI) refer to gauge pressure herein. As will be seen later, much of the hydraulic analysis of cooling water systems is performed in units of "head" in feet of water. For most cooling water systems, $1.0 \ lbf/in^2$ is equal to 2.31 ft of head.

2.3   DENSITY

The properties of water, whether in the form of ice, liquid, or steam, depend on the state of the water and are independent of the path that was taken to achieve the state. For example, the properties (e.g. temperature, pressure, etc.) of a volume of liquid water are independent of whether it was condensed in a cloud or melted from ice. Properties may be either extensive or intensive. For example, a bucket of water having a mass (the amount of matter contained in a substance), volume, pressure, and temperature might be divided equally into two buckets. The water in each bucket might still have the same pressure and temperature, but the mass and volume in each bucket would be one half of that contained in the original bucket. The mass and volume are said to be extensive properties, while the pressure and temperature are said to be intensive properties, because they are the same for each small element of water in the bucket. However, as shown in Equation 2.2, if one divides the mass by the volume, one gets density (an intensive property).

$$\rho = \frac{m}{V} \qquad (2.2)$$

where

ρ = density
m = mass
V = volume.

2.4 VISCOSITY

The viscosity of water flowing in a pipe is a measure of its resistance to flow. There are two measures of viscosity: dynamic (or absolute) and kinematic. Water is said to be a "Newtonian" fluid, meaning that its dynamic viscosity is a function of temperature alone. Dynamic viscosity is used to calculate the Reynolds number and Prandtl number of water. Kinematic viscosity is simply the dynamic viscosity divided by the density of the water. Dynamic viscosity is an indication of the pressure drop required to pass the water through a pipe, and kinematic viscosity is an indication of how fast the water is moving when that pressure is applied.

2.5 REYOLDS NUMBER

The Reynolds number of flow in a pipe is a non-dimensional property characterizing the flow as laminar or turbulent. The Reynolds number depends on the liquid velocity, the viscosity, and the pipe diameter and is calculated as follows:

$$\text{Re} = \frac{V_p \, d_i}{\mu} \qquad (2.3)$$

where
V_p = water velocity in a pipe or tube
d_i = inside diameter of a pipe or tube
μ = dynamic viscosity of the water.

Figure 2.4 shows the Moody diagram relating the Reynolds number to the friction factor.[1] If the Reynolds number is less that 2000, the flow is said to be laminar. If the number is greater than 4000, it is said to be turbulent in the pipe or tube. Between

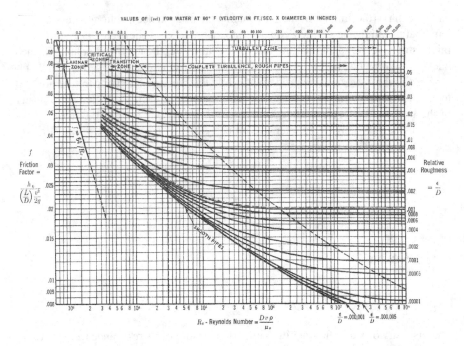

FIGURE 2.4 Moody diagram.

these values, transitional flow exists. If one knew the relative roughness of the pipe, one could calculate the pressure drop. Unfortunately, for CS cooling water piping, determining the relative roughness is a function of the tuberculation that frequently occurs in the pipe due to MIC, etc. In that case, the only major benefit derived from the Moody diagram is to verify that the flow is turbulent. Although in piping systems the flow is almost always turbulent, in some HX applications the flow may be laminar, affecting both the pressure drop through the HX and the rate of heat transfer.

2.6 PRANDTL NUMBER

The Prandtl number is a key parameter in determining heat transfer in a HX. The Prandtl number is defined as the ratio of the momentum diffusivity to the thermal diffusivity and depends on the dynamic viscosity, specific heat, and thermal conductivity of the water and is calculated as follows:

$$\text{Pr} = \frac{\mu\ c_p}{k} \tag{2.4}$$

where

Pr = Prandtl number
c_p = specific heat
k = thermal conductivity of water.

2.7   SPECIFIC GRAVITY

The specific gravity of water is the ratio of the density of water at a given temperature to that of water at a reference temperature. In English units, the reference temperature is 60°F (15.6°C), and the density of water is 62.36 lb/ft³ (1,000 kg/m³).

REFERENCE

1. Moody, L. F., Friction Factors for Pipe Flow. *ASME Transactions*, Vol. 66, pp. 671–684, 1944.

3 Intake Structures

3.1 INTAKE STRUCTURE DESIGN

The intake structure is a very important component of a cooling water system. Good engineering practice in the design of vertical turbine mixed flow or axial flow pumps in rectangular wet-pit intake structures is well established as a function of the pump bell diameter, D, by the Hydraulic Institute Standards.[1] Figure 3.1 shows a typical rectangular wet-pit intake structure. The pump bay width, W, and depth, H, should be sufficient to limit the maximum pump approach velocity to 1.0 to 1.5 ft/sec (0.30 to 0.46 m/s).[1,2] The minimum submergence, S, required to avoid vortices is a function of the dimensionless Froude number defined as follows[1]:

$$F_D = \frac{V_S}{\sqrt{gD}} \tag{3.1}$$

where
  F_D = Froude number
  V_S = velocity at suction inlet
  D = pump bell diameter
  g = gravitational constant

The minimum submergence, S, is as follows:

$$S_{min} = D\left(1 + 2.3F_D\right)$$

where the units must be consistent to yield a dimensionless Froude number and S is in the same units as D.

Although among pump types and manufacturers there may be some variation in the bell velocity (i.e. flow rate divided by the pump bell area), this is of secondary importance, since basing the dimensions on the pump bell diameter would ensure geometric similarity of flow patterns.[1]

The intake should supply an evenly distributed flow of water to the suction bell, as an uneven distribution of flow may result in vortices that may introduce air into the pump, resulting in reduced pump capacity.[1] The amount of submergence for successful operation depends upon the design of the approach to the pump intake and the size of the pump.[1] Dimension "H" is based on "minimum normal water level", taking into consideration normal operating friction losses through intake channel, trash rack, and the traveling screen.[1]

If the bottom elevation of the sump must be below the source of the cooling water to meet submergence requirements, the slope of the floor of the intake structure should not be greater than 10 degrees.[1]

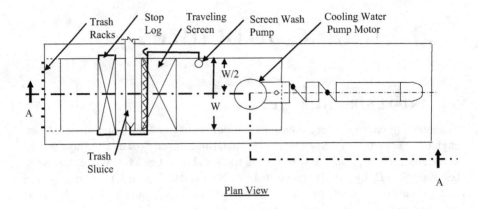

Plan View

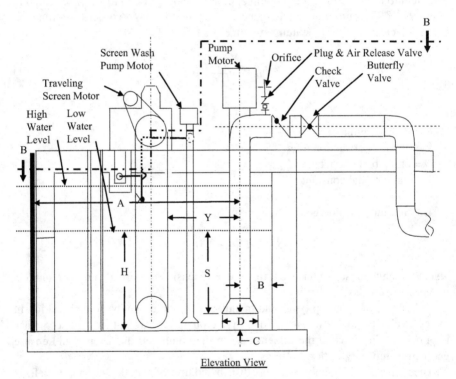

Elevation View

FIGURE 3.1   Intake pumping station.

The recommended values dimensions A, B, C, H, S, and Y as a function of "D" are shown in Table 3.1 from Reference 1.

It is noted that Figure 1 in Reference 2 recommends B = 0.8D and C = 0.5D.

Of course the IPS frequently must be designed long before the cooling water pump is selected. In that case, Figure 81 of Reference 3 may be consulted, where the sump dimensions are a function of the pump flow on a log-log scale.[3] Table 3.2 indicates some examples of approximate dimensions from Reference 3.

TABLE 3.1
Sump Dimensions as a Function of Pump Bell Diameter

Dimension	Number of Diameters
A	5D
B	0.75D
C	0.3D to 0.5D
H	S+C
S	See Equation (3.1)
Y	4D

TABLE 3.2
Sump Dimensions vs. Pump Flow

Dimension	Pump Flow (gal/min)	Sump Dimension (in)	Pump Flow (gal/min)	Sump Dimension (in)
A	3,000	87	10,000	165
B	3,000	14	10,000	26
C	3,000	13	10,000	19
H	3,000	41	10,000	58
W*	3,000	34	10,000	61
Y	3,000	60	10,000	108

* In Figure 81 of Reference 3, the sump width is noted as S, not W.

Assuming a pump bell velocity of 5.0 ft/sec (1.524 m/s), the sump dimensions in Table 3.2 are in good agreement with those calculated from Table 3.1 with the exception of the value for C, which is considerably higher than that calculated from Table 3.1 for 3,000 and 10,000 gal.min pumps (208% and 166%, respectively).

Of course many if not most IPS are considerably more complex than that shown in Figure 3.1. In many cases, a single IPS houses both the CCW pumps and the cooling water pump and may also house high pressure fire protection pumps and service water pumps. Figure 3.2 illustrates such an IPS.

Preferably water should not flow past one pump to reach the next pump.[3] If such an arrangement is required, dividing walls should be located around each pump to provide conditions similar to those in Figure 3.1, and/or turning vanes may be required under the pump to deflect the water upward.[3]

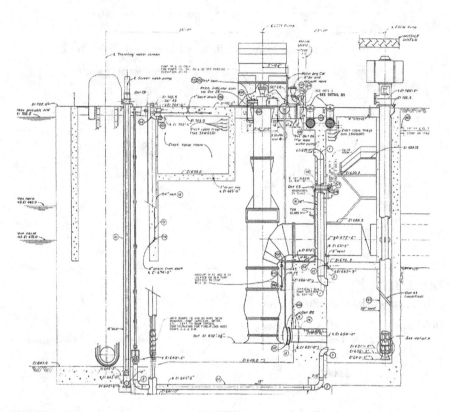

FIGURE 3.2   Intake pumping station with CCW pumps and cooling water pumps.

3.2   INTAKE STRUCTURE DESIGN DEFICIENCIES

As documented in Section 1.3, the failure of IPS to effectively protect cooling water systems from the intrusion of all manner of debris can be a serious problem. In 2007, the Institute of Nuclear Power Operations (INPO) issued a topical report on intake structure blockage vulnerabilities indicating that these events were continuing to happen. Many events result in power reduction or unit shutdown. Therefore, notwithstanding the impact on nuclear safety, a business case may be made for enhancing the design of the IPS to minimize the likelihood and the consequences of a blockage.

In many instances due to personnel turnover, power plant operators are unaware of prior events such as fish runs, seaweed, etc., that have challenged the cooling water system in the past. Operators are forced to rely on periodic walk-downs to detect such serious problems. When these events occur unexpectedly, the trash racks and traveling water screens can be quickly overwhelmed, leading to structural failure during environmental events such as storms. In very cold areas, frazil ice can quickly form on the screens. In other areas, oil spills can quickly coat the HXs served by the cooling water system with a film that degrades their heat transfer capability.

The screening material in many existing traveling water screens are of CS that has been in service for decades. The ability of these screens to withstand the forces resulting from a significant challenge is questionable.

3.3   REQUIREMENTS OF §316(B) OF THE CLEAN WATER ACT

Section 316(b) of the Clean Water Act (CWA) requires power and industrial plants that draw water from the waters of the United States to implement the "best technology available" (BTA) to reduce injury or death of fish and other aquatic life that may be killed by being impinged on or entrained in the IPS and subsequently killed by being exposed to the temperature and pressure changes associated with the CCW system and/or the cooling water system as they pass through the plant. The Environmental Protection Agency (EPA) estimates that the Act will affect approximately 670 U.S. power plants.[4]

The final EPA Rule promulgated on May 19, 2014, required new plants at existing facilities to install closed-cycle recirculating systems or to reduce actual intake flow to a level commensurate with that attained by a closed-cycle recirculating system. Another alternative would be to require plants to demonstrate that they have installed and will operate and maintain technological or other control measures that reduce the level of adverse environmental impact to a level comparable to that achieved through flow reductions commensurate with the use of a closed-cycle recirculating system.[5]

Existing plants with intake flows of over 2 million gallons of water per day that use 25% of the water for cooling are subject to the rule, and they must implement one of the following alternatives which may be deemed to be the BTA:

1. Install a closed-cycle recirculating heat rejection system;
2. Install an IPS with a maximum through-screen design intake velocity of 0.5 ft/ sec (0.15 m/s);
3. Operate an IPS with a maximum through-screen intake velocity of 0.5 ft/sec (0.15 m/s);
4. Install an intake at least 800 ft (244 m) offshore with a velocity cap;
5. Install traveling water screens that the EPA determines meet the BTA;
6. Implement any combination of measures that the EPA determines meets the BTA;
7. Achieve impingement mortality reduction to no more than 24%.[5]

These requirements are implemented through the system of required National Pollution Discharge Elimination System (NPDES) permits.[5]

Alternatives 1, 2, 3, and 4 were pre-approved by the EPA as meeting BTA. The other alternatives require proof that they are the BTA. The first and most expensive alternative would require the utility to convert the CCW system from an open system drawing water from a lake, river, or ocean to a closed system in which cooling towers would be added to the plant, thus reducing the intake requirement to only that required to replace the water evaporated from the system and the blowdown required to control the concentration of dissolved solids in the CCW system. The fifth option, commonly proposed, is to modify or replace the existing traveling water screens with screens designed to minimize fish impingement onto the screens. However, the Rule specified that if the NPDES permit director determines that the modified screens are insufficient, additional measures may be required. This option may pose greater challenges for the reliability of the cooling water system. In a report dated November 2001, the EPA reviewed the efficacy of alternatives to conventional traveling screens for cooling water intake structures including the following:

- modified traveling screens and fish handling and return systems
- cylindrical wedgewire screens
- fine mesh screens
- fish net barriers
- velocity caps
- aquatic micro-filtration barriers
- louver systems
- angled and modular inclined screens
- porous dikes and leaky dams
- behavioral systems.[6]

The EPA reported that conventional traveling screens used by approximately 60% of power plants have 3/8-inch (0.9525 cm) mesh wire to prevent clogging of HX tubes with screen wash at a typical pressure between 80 and 120 lbf/in (552 and 827 kPa).[6] These screens are normally rotated and washed only intermittently, so fish that are impinged on the screens by high intake velocity for extended periods of time die there or are killed by the high-pressure wash when the screens are operating.[5]

Ristroph screens are conventional traveling screens that have been modified with baskets or buckets on the front of the screens as shown in Figure 3.3 to hold fish in the water until the screen rotates to a point where the fish are spilled onto a trough where they are sluiced along with any trash back into the source water.[6] Of course, the screens should be continuously operating when large numbers of fish are present.

FIGURE 3.3   Ristroph traveling screen.

Operating experience at a number of power plants has shown at least a 70-80% reduction in impingement can be achieved over conventional traveling screens.[5]

Cylindrical wedgewire screens like the one shown in Figure 3.4 have been shown to eliminate impingement at Eddystone Generation Station.[6] In this design, ambient flow is in the axial direction of the screen. As the aquatic organisms are carried past the screen, they are restricted from entering by their inability to fit through the screen spacing. Their momentum carries them downstream as their ability to actively maneuver away from the screen's hydraulic zone of influence aids in their passage.[7] However, this technology has not been widely utilized in other power plants.[6]

One obvious solution to eliminate entrainment of fish eggs, larvae, and juvenile fish would be to mount fine mesh screens on conventional traveling water screens. This method was implemented at the Big Bend and Brunswick Power Plants, suggesting that fine mesh screens can reduce entrainment by 80% or more but require intensive maintenance when they are in use.[6] Additional full-scale performance data on fine mesh screens were not available at the time of publication of Reference 6.[6]

Fish net barriers are wide-mesh nets placed in the source water attached to a floating boom in front of the entrance to intake structures where relatively low entrance velocities exist. The mesh is sized to prevent the indigenous fish from passing through the net. Fish net barriers have been used at numerous facilities where they have proven effective at reducing impingement by over 80%.[6] However, experience shows that high debris flows can cause significant damage to the nets, necessitating frequent maintenance.

When a velocity cap is placed over a vertical inlet at offshore intakes, it converts the vertical flow to horizontal flow which fish can more easily avoid, thus reducing impingement.[6] When the intake is extended to an area of deeper, cooler, water with less biological density, the entrainment of aquatic organisms may also be reduced.[7] (Note that installation of an intake at least 800 ft (244 m) offshore with a velocity cap constitute BTA.)

Other than those technologies discussed above, none of the other technologies described in Reference 6 had been implemented full-scale in power plants when the document was published.

In 2014, Bowman proposed an alternative for reducing both impingement and entrainment not previously considered by the EPA or others.[8] The CCW systems of many power plants were originally designed to take advantage of the abundant and

FIGURE 3.4 Cylindrical wedgewire screen.

free CCW that was available from rivers, lakes, and oceans. The economic optimum design was a high-flow, low head, system with a single-pressure MC designed with minimal surface areas. Therefore, the potential may exist to reduce the required CCW flow to existing power plants by redesigning and modifying the existing CCW system to minimize intake flow based on current technology. The result could be a new and improved MC and other turbine cycle equipment and perhaps new CCW pumps and/or turbine rotors, resulting in the same or better plant performance. The following is taken from Reference 9 where "NDCT" stands for "Natural Draft Cooling Tower".[9]

"Alternative C – Refurbish the existing two NDCT and convert the existing MC into a multi-pressure condenser.

> Alternative C would also require the refurbishment of the existing two NDCTs. However, in lieu of an expensive third NDCT, the existing MC would be reconfigured such that the CCW flow through the three shells of the MC would be in series rather than in parallel. The existing 10 foot square pressure equalizing ducts between the three shells would be blocked, making it a multi-pressure condenser. The CCW flow through the MC would be reduced by one-third and the velocity through the 0.75-inch (1.91 cm) titanium tubes would be increased from 5.5 to 10.9 ft/sec. To accommodate the higher pressure drop through the MC tubes, the CCW Pumps would be replaced with higher head pumps and motors. A new transformer would be provided to serve the larger motors. To accommodate the higher head pumps, 10 foot diameter cement mortar-lined CS piping would be installed inside of the existing 13.5' × 13.5' concrete conduits between the IPS and the MC. By reducing the CCW flow rate by one-third, the water loading on the existing NDCT would be reduced. Even though the temperature rise through the MC would be increased and the hot water temperature would be greater, the cold water temperature coming from the NDCT would be less. For a NDCT, the hot water temperature has only a small effect on the cold water temperature, as a higher hot water temperature increases the air flow through the tower. The overall reduction in electrical output of this alternative is 41 MWe per unit per year or 3.5%.

A reduction on the CCW flow by one-third constitutes a significant reduction on both impingement and entrainment and in some instances may reduce the intake velocity to below 0.5 ft/sec (0.15 m/s), which is BTA or in other instances taken with other measures described herein may also result in BTA. For example, a reduction of CCW flow by one-third may reduce the cost of adding cooling towers by a similar amount.

3.4 POTENTIAL IMPROVEMENTS IN THE ENGINEERING OF INTAKE PUMPING STATIONS

Since one of the significant challenges to the free flow of cooling water is the intrusion of aquatic animals, some of the technologies designed to protect aquatic life also enhance cooling water system reliability. Offshore intakes with a velocity cap not only reduce the intake of aquatic life but also that of algae and marsh grass. Floating booms used to support fish nets also protect against oil slicks and security challenges.

Skimmer walls across the intake channel that restrict intake flow to the bottom of the channel can achieve similar results at some locations. Continuous operation of a traveling water screen as required with fish baskets or buckets protects the cooling water system from unexpected events that can quickly overcome the intake.

Figure 3.5 illustrates several modifications to the wet-pit intake structure shown in Figure 3.1. The following enhancements are proposed:

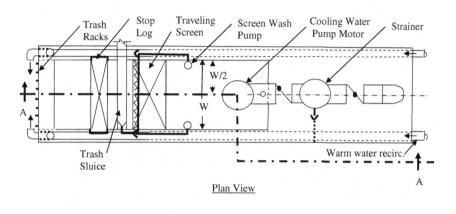

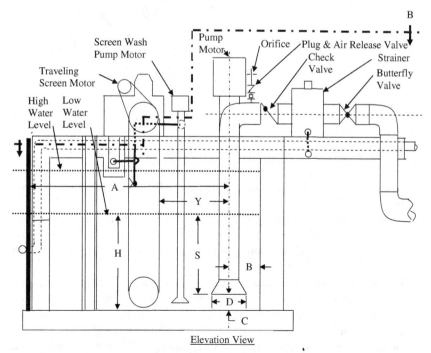

FIGURE 3.5 Enhanced intake pumping station.

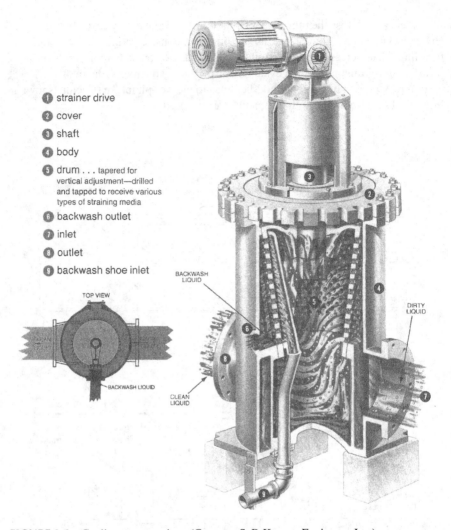

① strainer drive

② cover

③ shaft

④ body

⑤ drum . . . tapered for vertical adjustment—drilled and tapped to receive various types of straining media

⑥ backwash outlet

⑦ inlet

⑧ outlet

⑨ backwash shoe inlet

FIGURE 3.6   Cooling water strainer. (Courtesy S. P. Kenney Engineers, Inc.)

- In areas where frazil ice is a hazard, design the IPS to accept recirculated warm water from the condenser discharge;
- Provide redundant screen wash pumps so that the cooling water system is not vulnerable to challenges when one pump or motor is out of service and so that the second screen wash pump may be started to increase screen wash flow if required during an intrusion;
- Provide automatic backwashing cooling water strainers at the discharge of the cooling water pumps to strain out all but larvae, veligers (larval with ciliated flaps for swimming) and fish eggs, etc.;
- Increase cooling water pump submergence such that the pumps would continue to operate satisfactorily even when the structural design maximum differential pressure across the traveling screens is reached;

- Provide an automatic start on high differential pressure across the screen if traveling water screens are not continuously operated (recommended);
- Employ screens made of only SS mesh;
- Provide cameras to permit the operator to observe the condition of the traveling screen and trash rack;
- Use anti-foulant paint on the interior walls of the wetted surface of the IPS if located in an area where zebra mussels are prevalent.

As discussed in Chapters 6 and 7, straining the cooling water at the discharge of the cooling water pump is critical for controlling biofouling and MIC. There are two types of large automatic backwashing strainers: those that utilize wedgewire and those that utilize individual baskets that screw into a drum like the one shown in Figure 3.6 above which are recommended.

The pressure drop across a 24″ (61.0 cm) S. P. Kinney Model A strainer with wire baskets rated at 16,000 gal/min (1,000 l/s) capacity was measured during the preoperational testing of the safety-related cooling water system at the Sequoyah Nuclear Plant (SNP). This test indicated a pressure drop across the strainer of 5.0 lbf/in (34.5 kPa) which was several times more than that advertised by the strainer manufacturer.

REFERENCES

1. *American National Standard for Pump Intake Design*, ANSI/HI 9.8-1998, Hydraulic Institute, Parsippany, NJ, 1998.
2. Greutink, Herman. Recommendations for Vertical Pump Intakes, *Pumps and Systems Magazine*, Birmingham, AL, November, 1994.
3. *Hydraulic Institute Standards for Centrifugal, Rotary, and Reciprocating Pumps*, 14th ed., Hydraulic Institute, Parsippany, NJ, 1983.
4. Russell, Ray. EPA's 316(b) Rule: Are You Ready? *Power Engineering*, September 2013.
5. 40 CFR Part 122 and 125, National Pollution Discharge Elimination – Final Regulations to Establish Requirements for Cooling Water Intake Structures at Existing Facilities and Amend Requirements at Phase I Facilities, Environmental Protection Agency, Washington, DC, May 9, 2014.
6. *Technical Development Document for the Final Regulations Addressing Cooling Water Intake Structures for New Facilities*, EPA-821-R-01-036, Chapter 5, Environmental Protection Agency, Washington, DC, November, 2001.
7. Clubb, Richard. When Budgeting for §316(b) Compliance, Consider All Options. *Power Engineering Magazine*, Tulsa, OK, March 2013.
8. Bowman, C. F. *The Third Option for Meeting 316(b) Requirements, Power 2014-32113, Proceedings of the ASME 2014 Power Conference*, Baltimore, MD, July 2014.
9. Sequoyah Nuclear Plant §316(b) – §122.21(r)(10) – (13) Information, Appendix A to Transmittal from TVA to Tennessee Department of Environment and Conservation, June 29, 2018.

4 Pumps

4.1 TYPES OF PUMPS

Cooling water and CCW system pumps may be classified as either vertical wet pit pumps as described in Chapter 3 or dry-pit pumps as discussed in Section 1.2. In power plants with open once-through CCW systems, the CCW pumps are typically of the vertical wet pit type as shown in Figure 3.2. In some power plants with cooling towers, the CCW pumps may be arranged as dry-pit pumps with the shaft and motor mounted either horizontally or vertically as shown in Figure 13 in Reference 1. Cooling water system pumps that are supplied from a lake, river, or ocean, etc., as shown in Figures 3.1 and 3.5 are of the vertical wet pit type. However, in some power plants, the cooling water pumps take suction from the CCW pumps and are horizontal pumps located inside the turbine building.

4.2 TYPES OF PUMP IMPELLERS

Pumps may be classified as having radial flow, mixed flow, or axial flow impellers. Pumps having radial flow or mixed flow impellers are referred to as centrifugal pumps, because they employ centrifugal force to sling the water in a direction perpendicular or almost perpendicular to the pump shaft, whereas pumps with axial flow impellers (referred to as propeller pumps) move the flow along the axis of the pump.[1] In power plants with open once-through CCW systems, the CCW pumps are typically low-head axial flow pumps. Since a siphon is established through the MC, the CCW is only required to overcome the friction head losses through the system. In power plants with closed CCW systems with cooling towers, the CCW pumps are typically mixed flow pumps, because the CCW pump must overcome the static head required to return the CCW to the cooling tower. Vertical wet pit cooling water pumps are typically mixed flow pumps, whereas dry-pit pumps are typically radial-flow centrifugal pumps.

A vertical wet pit pump impeller is housed in the pump bowl at the bottom of the line shaft that transmits the torque required to drive the pump from the motor above. Although not common in the case of cooling water pumps, there may be more than one pumping stage with multiple bowls and impellers stacked on top of each other with diffuser vanes in between to direct the flow from one stage to the next. A suction bell is located below the first stage bowl to direct the cooling water smoothly into the impeller. Replaceable bronze wear rings positioned between the impeller and the bowl minimize wear of both the impeller and the bowl casing and enhance efficiency.

Figures 4.1 and 4.2 illustrate the two kinds of mixed flow impellers: semi-open (Figure 4.1) and closed (Figure 4.2). Experts do not agree as to which design is more efficient. Both have their advantages and disadvantages. With a semi-open impeller, the impeller rides very close to the wear ring below, and this "endplay" or the amount of vertical clearance between the impeller and the wear ring requires proper

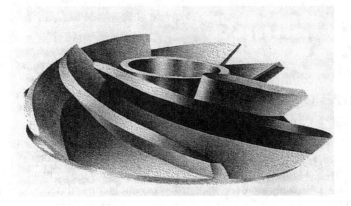

FIGURE 4.1   Semi-open impeller.

FIGURE 4.2   Enclosed impeller.

adjustment. Initially, the impeller rests on this seat, and the clearance is adjusted to allow for the line shaft to stretch while maintaining a close clearance during operation. The impeller clearance is adjusted by means of an adjusting nut at the top of the motor. However, if the line shaft and the pump column are of different materials, this clearance can easily come out of adjustment due to cooling water temperature changes. By contrast, the wear ring encloses a closed impeller in the vertical direction which is free to move up and down slightly due to changes in the cooling water temperature as the line shaft expands or contracts relative to the pump casing. However, whereas the clearance between a semi-open impeller may be adjusted to compensate for wear, the same is not true with a closed impeller.

Figure 4.3 shows a cooling water pump with a semi-open impeller on the left and a CCW pump with a propeller on the right.

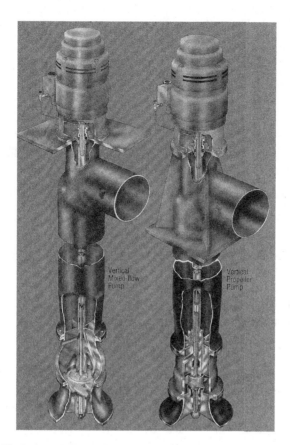

FIGURE 4.3 Vertical wet pit cooling water and CCW pumps. (Courtesy of Johnston Pump Co.)

4.3 SPECIFIC SPEED

The specific speed of an impeller, n, is defined as the revolutions per minute at which a geometrically similar impeller would run if it were of such a size as to discharge one gallon per minute against one foot of head. Specific speed is a parameter used to select the type of pump that is best suited to a particular application and to check suction limitations of the pump.[2,3] The shape of the pump impeller producing maximum efficiency is a function of the specific speed as defined by Equation (4.1).[1]

$$n = \frac{NQ^{\frac{1}{2}}}{H^{\frac{3}{4}}}$$

(4.1)

where
N = rototional speed, rev/min
Q = flow rate, gal/min
H = total head, ft.

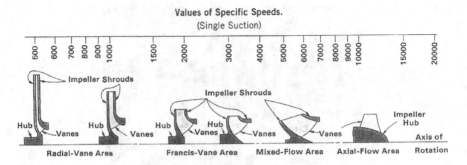

FIGURE 4.4   Pump impeller profile as a function of specific speed.

Figure 4.4 shows the various pump impeller profiles as a function of specific speed.

The following shows the type of impeller that would yield the best efficiency for the indicated range of specific speeds in English units[1]:

500–3,500 radial flow

3,500–7,500 mixed flow

7,500–12,000 axial flow.

Typical cooling water pumps with flow rates of from 1,000 to 10,000 gal/min would achieve maximum pump efficiency at a specific speed of approximately 3,000.[1,2] A centrifugal pump impeller may be replaced with a smaller or larger impeller in the same bowl size, within limits.[4]

4.4   CAVITATION AND PUMP SUCTION SPECIFIC SPEED

Cavitation occurs in pumps when the absolute pressure at the pump inlet decreases below the vapor pressure of the fluid at which time vapor bubbles form at the impeller inlet.[3] When the bubbles pass through the impeller and enter the higher pressure region of the pump, they collapse. When that collapse occurs on the surface of the impeller, the liquid rapidly moves in to fill the space left by the bubble, impacting a small area of the impeller with a very high localized pressure surge that can pit and erode the impeller.[3]

Suction specific speed, s, is an index number that describes the characteristics of the suction of a vertical wet pit pump and is defined in Equation (4.2).[2]

$$s = \frac{NQ^{\frac{1}{2}}}{(NPSHA)^{\frac{3}{4}}} \geq \frac{NQ^{\frac{1}{2}}}{(NPSHR)^{\frac{3}{4}}} \qquad (4.2)$$

where
  $NPSHA$ = net positive suction head available, ft.
  $NPSHR$ = net positive suction head required, ft.

The available value of "*s*" must equal or exceed that required value in order to prevent cavitation. A high value of "*s*" at the operating point of maximum efficiency indicates a good suction design.

For a vertical wet pit pump, NPSHA may be calculated as in Equation (4.3)

$$NPSHA = P_b + S - P_v \tag{4.3}$$

where

P_b = barometric pressure
S = submergence
P_v = vapor pressure at the cooling water temperature.

NPSHR is dictated by the pump design and is specified by the pump manufacturer. (See Figure 4.5 below.) A reasonable suction specific speed is 8,500 rev/min.[2]

4.5   SYSTEM HEAD CURVE

The dashed line in Figure 4.5 illustrates the system head curve for a cooling water system. In this example, the static head of the system is indicated by the Y intercept at zero flow rate. As flow increases, the required head to force the flow through the system also increases by a ratio of the flow to the 1.8 to 2.0 power. Since most cooling water systems serve more than one HX, the pressure drop through the various parallel loops in the system is balanced to the extent possible to achieve the desired flow to each component by judicious selection of pipe sizes. (See Chapter 7.) However, the required pump head to achieve the desired total flow rate through the system is normally based on the limiting flow path, and the flow to other components is set by throttling control valves as required to achieve the desired flow balance. (See Chapter 8.)

4.6   PUMP HEAD CAPACITY CURVE

The pump head-capacity curve in Figure 4.5 shows how much flow a particular cooling water pump can deliver as a function of the required system head. The same pump may meet the flow requirements for a wide variety of system head curves, but as the flow demand increases, the available head decreases. The point at which the pump head curve intersects the Y axis (i.e. zero flow) is known as "shut-off head". Although a continuously rising characteristic pump curve as it approaches shut-off head is desirable for controllability, limiting the pump head at shut-off may be desirable to limit the system design pressure. At the other extreme, the maximum pump flow is known as the pump "run-out", which is limited, since pump cavitation may occur when the pump operating head is much lower than the rated head and the NPSHR increases exponentially.

Figure 4.6 shows the pump head-capacity curve for the CCW pump shown in Figure 3.2. The absence of a continuously rising head curve is not uncommon among CCW pumps. When this is the case, the point of normal operation should be well to the right of the hump in the curve for stable operation.

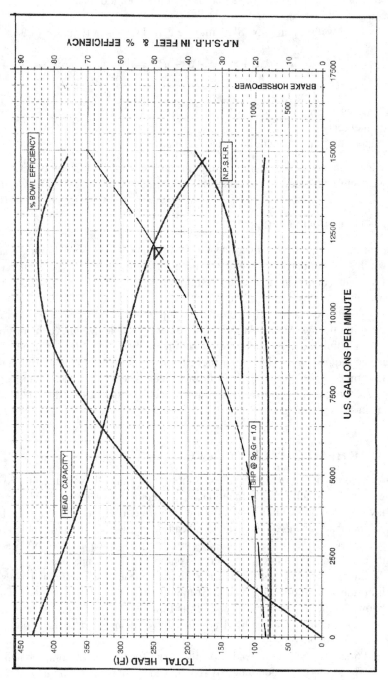

FIGURE 4.5   System and pump curves. (Courtesy Sulzer Pump Co.)

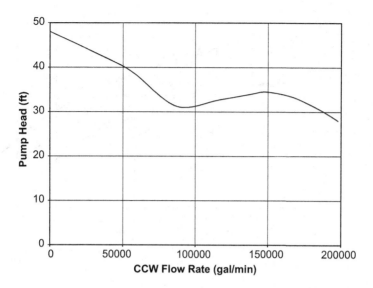

FIGURE 4.6 CCW pump head-capacity curve.

4.7 PUMP SIZE, EFFICIENCY, AND SPEED

The impeller of a pump is configured to achieve maximum efficiency at a single point along the pump head-capacity curve where there is minimum radial load on the impeller and minimum cavitation.[3] As one may see from Figure 4.5, pump selection is based on the point of normal pump operating being at this point of maximum efficiency. Normally, the efficiencies of centrifugal pumps vary between 50% and 85% with larger pumps being more efficient than smaller ones as illustrated by Figure 4.7. The capacity of a centrifugal pump is proportional to its rotational speed, which normally can vary from between 800 and 3,000 rev/min, depending upon the design and rated capacity of the pump. Cooling water pumps normally operate at a constant speed during operation. Pumps that operate at slower speeds are more efficient than those that operate at higher speeds. The specific speed that is available according to Equation (4.2) is a valuable criterion in determining the allowable maximum speed of the pump.[2]

The brake horsepower, *BHP*, required to drive a pump in English units is defined in Equation (4.4).[5]

$$BHP = \frac{Q\,H\,\,s.g.}{3960\,\,\eta_p} \tag{4.4}$$

where
$s.g.$ = specific gravity of the cooling water
η_p = pump efficiency.

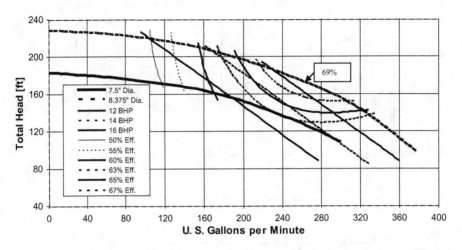

FIGURE 4.7   Pump head and efficiency curves. (Courtesy FLOWSERVE Pump Division).

4.8   NET POSITIVE SUCTION HEAD

As discussed in Section 4.4, in order to prevent cavitation occurring at the pump impeller, *NPSHA* must equal or exceed *NPSHR*. *NPSHR* is dictated by the pump design, and the pump manufacturer may be able within limits to engineer a pump with lower values of *NPSHR*.

The "Thoma's cavitation constant", σ, defined as the ratio of the net positive suction head at the point of cavitation inception, $NPSH_i$, to H may be calculated from Equation (4.5).[3]

$$\sigma = \frac{NPSH_i}{H} = \frac{Kn^{4/3}}{10^6} \qquad (4.5)$$

where values of K are given in Table 4.1.[3]

As one may see from Equation (4.5) and Table 4.1, $NPSH_i$ increases with H and n and decreases with increased η_p. When applied to multistage pumps, H is the total head per stage, so one way to reduce $NPSH_i$ would be to design a very efficient pump with two stages.

TABLE 4.1
Values of *K* for Equation (4.5)[3]

Pump Efficiency (%)	SI Units	English Units
70	1726	9.4
80	1210	6.3
90	796	4.3

4.9 PUMP MINIMUM FLOW RATE

As one may see from Figure 4.5, even at shut-off head the pump motor brake horse-power transmitted to pump can be quite high, due to poor pump efficiency at that point. The pump is cooled by the cooling water flow through the pump. If the flow through the pump is too low, the pump can overheat. As a general rule, the minimum flow through a cooling water pump should be no less than 40% of the flow rate at the point of maximum efficiency. Large high horsepower pumps may have a minimum flow limitation as high as 70% of the flow rate at the point of maximum efficiency.[6] Since in the case of cooling water systems the required flow rate may vary with cooling water temperature, multiple pumps operating in parallel may be required to not only avoid overheating the pump but also to be able to operate at closer to the point of maximum efficiency when the temperature of the cooling water is low.

4.10 AIR VENTING

As shown in Figures 3.1 and 3.5, a check valve is normally located at the discharge of a wet pit vertical turbine-type cooling water pump to prevent the entire cooling water system from draining down each time a pump is stopped. When the pump stops, normal practice is to bleed air into the pump column through a float-type air release valve or through an inverted check valve as shown in the figures. When the pump is restarted, the air in the pump column is gradually released through this valve to control the rise of water in the pump column to avoid a water hammer as the column of water hits the closed check valve. (See Chapter 8.) Care must be taken in sizing the air release mechanism, since if it vents the air too slowly, the remaining air will pass through the discharge check valve with the water, and if too quickly, a water hammer may occur. Although a float-type air release valve or inverted check valve of suitable size may be available, most likely the correct air flow would be dictated by an appropriately sized orifice as shown in the figures. The mass flow rate of air may be determined by the volumetric flow rate of the pump at rated conditions times the density of air. Equation (4.6) from Page 3–24 of Reference 7 shows the relationship between the mass flow rate of air as a function of the diameter of the orifice in English units.

$$w = 0.525 \ Y \ d_1^2 \ C \ \sqrt{\Delta P \ \rho_1} \qquad (4.6)$$

where
w = mass flow rate of air, lbm/sec
Y = expansion factor (See Page A-21 of Reference 7)
d_1 = orifice diameter, inches
C = orifice flow coefficient
ΔP = pressure drop across the orifice, lbf/in^2.

The size of the piping and valves up stream of the orifice should be such that the pressure drop through them should be no more than approximately 10% of the pressure drop through the orifice.

In an open CCW system, the siphon through the condenser is normally maintained when the CCW pumps are stopped. Refilling the system and reestablishing condenser vacuum is a long and laborious process, so the system would not be allowed to drain down by continuing to operate the condenser vacuum priming pumps. The condenser vacuum is maintained by continuing to operate the condenser vacuum pumps. If it becomes necessary to allow the water to drain from the MC, it would be refilled by throttling the condenser discharge valves to raise the hydraulic gradient sufficient to fill the condenser tubes as air is vented through the vacuum priming system to control the rate at which the water rises in the tubes to avoid a water hammer. (See Chapter 8.)

4.11 PUMP LINE SHAFT SEALS

The line shaft connecting the impeller of a vertical wet pit pump may be either closed or open. The pumps in Figure 4.3 have a closed line shaft (i.e. the shaft is enclosed within a pipe). Pumps with long line shafts must have bearings supporting the shaft spaced so that the natural frequency is above that of the pump's rotational speed to prevent vibration. The bearings are either cutlass rubber or phenol plastic. If the shaft is enclosed, the lubricating water may come from down stream of the strainer shown in Figure 3.5 or from a separate seal water system. If the shaft is open, the bearing is lubricated by the cooling water passing by the bearing. Long line shafts require screwed, split ring, or flanged shaft couplings. Prior to pump start, prelubrication is required for the line shaft bearings, throttle bushing, and mechanical seal.

4.12 PUMP MOTORS

The required pump motor rated horsepower is defined in Equation (4.7).

$$MHP = \frac{BHP}{\eta_m} \tag{4.7}$$

where
MHP = motor horsepower
η_m = motor efficiency.

The motor efficiency is typically 95%.

The cooling water system pump motor should be sized so that the MHP requirement is not exceeded at any point over the normal operating range. The service factor of a motor is an indication of how much the nameplate rating of the motor may be exceeded for short periods of operation.

The type of motor depends on the function of the cooling water system. Safety-related cooling water pump motors are normally Nuclear Class 1E, totally enclosed, with a service factor of 1.0. Non-safety-related cooling water pump motors are normally National Electric Manufacturers Association MG1, open and drip-proof, but if the cooling water system is an outdoor installation, the pump motor should be weatherproof. The service factor for these pumps is normally 1.15. Motor cooling water lines (if required) should be self-draining to preclude freezing.

4.13   PUMP MARGIN AND IN-SERVICE TESTING

Under provisions of 10CFR50.55a, nuclear safety-related (essential) cooling water pumps are required to be inservice tested in accordance with Section XI of the American Society of Mechanical Engineers (ASME) Boiler and Pressure Vessel Code. ASME Section XI in turn invokes Reference 8. Subsections ISTB and ISTF of Reference 8 cover inservice testing of pumps in light-water reactor nuclear power plants before and after the year 2000, respectively. The test quantities are required to be measured and compared to reference values that are to be measured during initial operation or after refurbishment of the pump. The values to be measured include pump speed, differential pressure, flow rate, and vibration amplitude. The pump is to be operated at the nominal motor speed and the resistance of the system varied until the measured flow equals the reference point, whereupon the pump head is determined and compared with the reference value. Alternatively, the flow rate may be varied until the pump head matches the reference value and the flow rate is compared to the reference value. Note that the test is solely to determine the performance of the pump and does not speak to the flow-passing capability of the cooling water system. (See Chapter 7.)

The test acceptance criteria include an "alert range" and an "action range". If the results of the test fall within the former value, the frequency of the testing is to be doubled until the cause of the problem is determined and the condition corrected. If the results of the test fall within the latter value, the pump is to be declared inoperable.

Figure 4.8 shows the results of inservice testing of an essential cooling water pump. This pump was tested at a flow rate of 9,500 gal/min.

This pump had a semi-open impeller, and the line shaft was SS, while the pump casing was CS. In addition to the gradual declining performance of the pump, one can clearly see the impact of the seasonal change in the clearance between the impeller and the wear ring on performance as discussed in Section 4.2.

4.14   POTENTIAL IMPROVEMENTS IN THE ENGINEERING OF COOLING WATER PUMPS

A search of the INPO Nuclear Plant Reliability Data System over a four-year period revealed that the majority (72%) of the pump failures were due to external leakage which did not significantly affect operability. Of the remaining failures, approximately 40% of the failures were detected by periodic testing. The remainder were detected through operational abnormalities or while performing normal maintenance. Cooling water system pumps are particularly vulnerable to wear and failures due to the corrosive and abrasive nature of the raw water being pumped.

As illustrated by Figure 4.8, the performance of vertical mixed flow cooling water pumps can deteriorate over time. When the pump column and line shaft are of dissimilar metals, the seasonal loss on pumping head due to difference in thermal expansion can be significant if a semi-open impeller is employed. Although no such problem occurs with totally enclosed impellers, one does not have the ability to manually adjust the clearance between the impeller and the wearing ring without disassembling the pump in order to restore the operating clearance.

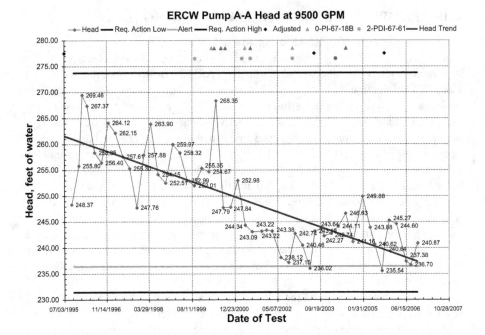

FIGURE 4.8   Inservice test results for a cooling water pump at a flow rate of 9,500 GPM.

The following enhancements to the engineering of cooling water pumps are proposed:

- The ability to test cooling water pumps is essential to diagnosing pump deterioration and correcting problems. Therefore, provisions should be made in the design of the cooling water system to be able to periodically test the pump. These provisions should include means to measure the flow rate and to calculate the total head of the pump and to provide a flow path capable of testing near the normal flow rate.
- Prelubrication of vertical wet pit mixed flow pumps should be provided from either down stream of the strainer shown in Figure 3.5 or from a separate seal water system.
- Small water lines such as those required to provide prelubrication and cooling water pump motor cooling should be protected by appropriate insulation and/or heat tracing.
- If a semi-open impeller is employed, the line shaft and the pump column should be of the same material. If the line shaft is of CS, disassembly can be very difficult due to corrosion. Therefore, both the line shaft and the pump column should be of SS.
- Considering the corrosive and erosive nature of RCW, serious consideration should be given to employing SS materials for the pump bowl, impeller, column, line shaft, and discharge head. (See Chapter 6.)

- Specify the appropriate heat treatment for each part to prevent stress corrosion cracking.
- For pumps in unfavorable sump configurations, consider employing features such as straightening vanes on the pump suction.

REFERENCES

1. Linsley, R. K. et al. *Water-Resources Engineering*, 4th ed., pp. 416–423, McGraw-Hill, Inc., 1992.
2. *Hydraulic Institute Standards for Centrifugal, Rotary, and Reciprocating Pumps*, 14th ed., p. 12, Hydraulic Institute, Parsippany, NJ, 1983.
3. Mays, L. W. *Water Resource Engineering*, 2nd ed., pp. 479–480, John Wiley & Sons, Inc., 2011.
4. Shashi, M. *Liquid Pipeline Hydraulics*, CED Engineering.
5. *Cameron Hydraulic Data*, 16th ed., pp. 1–25, Ingersoll-Rand, Woodcliff Lake, NJ, 1981.
6. Guyer, P. J. *Introduction to Pumping Stations for Water Supply Systems*, CED Engineering.
7. *Crane Technical Paper No. 410*, 24th Printing, Crane Co., King of Prussia, PA, 1988.
8. *Operation and Maintenance of Nuclear Plants*, ASME OM-2020, American Society of Mechanical Engineers, New York, NY, 2020.

5 Piping and Valves

5.1 PIPING APPLICATIONS

The purpose of cooling water piping is to transport water from one point in the system to another in an efficient and reliable manner for the life of the plant. The applications for cooling water piping may be broadly categorized as either buried pipe or pipe that is hung from pipe hangers inside a structure. A material that may be suitable for one application may not be suitable for the other.

5.2 PIPING MATERIALS

CS piping alone, though less expensive than some other alternatives, is generally a poor choice in the long run for either buried or hung applications when compared with other piping materials. As will be shown in Chapter 6, CS in cooling water systems is frequently not a suitable material to last for the life of the plant and must be replaced due to corrosion or flow blockage. As will be shown in Chapter 7, the flow-passing capability of CS piping in cooling water systems is often poor even in relatively new piping systems when compared with other materials. Indeed, some studies have shown that due to the tuberculation found in CS piping in cooling water systems, the line size required to pass the same flow rate with the same pressure drop may be one size smaller for corrosion-resistant materials than for CS, and the installed cost may be only a fraction more or less than for CS. CS pipe (typically ASTM A106 Grade B or A53 Grade B) has been used in all sizes up to more than 20 ft (6.1 m) in diameter. Steel pipe in sizes 0.5 to 12-inch (1.27 to 30.5 cm) in diameter is often a continuous tube formed by drawing over a mandrel. Larger steel pipe up to 48-inches (122 cm) is normally fabricated from long, narrow steel plate that is bent to shape and welded along a spiral joint.[1]

Suitable alternative materials, depending on where in the plant the piping is located, may include SS, CPP, pre-lined ductile iron (PDI) pipe, and a variety of plastic piping materials such as FRP, PVC, and HDPE pipe.

5.3 PIPE LININGS, COATINGS, AND WRAPS

Pipe linings and coatings including CS pipe lined with epoxy, PVC, SS, or cement mortar are available to the designer. Some linings and/or coatings may be more appropriate for replacement of existing CS piping, while others are best suited for new construction.

5.4 BURIED PIPING

A typical power plant has a large complement of cooling water piping buried in and around the facility. CS piping has been the material of choice for piping systems buried in the plant yard. Such piping is normally coated on the exterior with a bituminous coating and wrapped in fiberglass to protect against exterior corrosion. Depending on the composition of the soil surrounding the pipe, a cathodic protection system may be needed to supplement the exterior coating. As shown in Figure 5.1, cathodic protection consists of a direct current power source with the positive terminal connected to sacrificial anodes buried under ground. Other wires serving as cathodes are connected to the pipes. Any breach in the pipe coating provides a path for the electricity to pass from the anodes through the ground, back to the pipe, and back through the cable to the power source. The flow of electricity into the pipe limits the corrosion, thus protecting the pipe from loss of metal. Although frequently viewed as passive, cathodic protection systems require supervision. Excessive impressed voltage can cause disbanding of the pipe coating, resulting in an accelerated corrosion rate. Plant engineers are often not familiar with these systems, and when equipment fails (if it is discovered at all) it may be found to be obsolete and spare parts may not be available.

Other materials such as CPP, PDI pipe, and plastic piping may be more suitable for buried service. In recent years, there has been a trend toward using various plastic piping systems.

5.4.1 CONCRETE PRESSURE PIPES

The three types of CPP are reinforced, prestressed, and pretensioned concrete pipes. Reinforced concrete pipe consists of a thin steel cylinder with steel joint rings welded to its ends. The cylinder is surrounded with one or more cages of reinforcing made from bar, wire, or welded wire fabric. This assembly is encased in a wall of dense concrete covering the cylinder both inside and out. The pipes are manufactured according to American Water Works Association (AWWA) C300. Sizes covered by

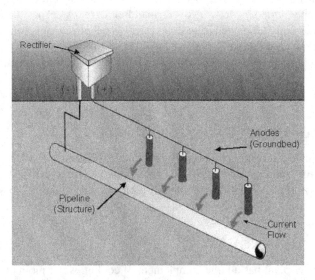

FIGURE 5.1   Cathodic protection of buried CS piping.

the standard are 20 through 90-inch (50.8 through 229 cm). Operating pressures can be as high as 260 lbf/in² (1,793 kPa). Standard lengths are 12, 16, and 20 feet (3.66, 4.88, and 6.10 m).[2]

Prestressed concrete pipe consists of a thin steel cylinder with steel joint rings welded to its ends. The cylinder is then lined with a dense coating of concrete and high tensile strength wire is wound around the outside of the cylinder at a predetermined stress. The pipes are manufactured according to AWWA C301. Sizes covered by the standard are 16 through 96-inch (40.6 through 244 cm). Operating pressure range for economical design is from 200 to 275 lbf/in² (1,379 to 1,896 kPa). Standard lengths are 16 and 20 feet (4.88 to 6.10 m).[2]

Pretensioned concrete pipe as shown in Figure 5.2 consists of a heavier steel cylinder with steel joint rings welded to its ends. The cylinder is then lined with a dense coating of concrete. High tensile strength wire is wound around the outside of the cylinder at a predetermined stress, and the outside is coated with concrete. The pipes are manufactured according to AWWA C303. Sizes covered by the standard are 10 through 36-inch (25.4 through 91.4 cm). Operating pressure range can be as high as 400 lbf/in².(2,758 kPa). Lengths range from 24 to 40 feet (7.32 to 12.2 m).[2]

There has been considerable operating experience with all three of these types of pipes, including some in nuclear seismic Category I applications. They are not subject to either interior or exterior corrosion. However, special consideration should be given to the joint details due to possible joint leakage.

5.4.2 Pre-Lined Ductile Iron Pipes

Pre-lined ductile iron pipe is produced in a factory by applying a high strength mortar to the interior of ductile iron pipe using either the centrifugal process or the projection method. In the former method, the mortar is distributed evenly throughout the

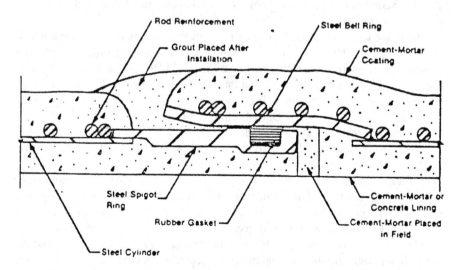

FIGURE 5.2 Pretensioned concrete pipe.

length of a section of pipe by moving a lance while the pipe is spinning. In the latter method, the mortar is sprayed or slung evenly onto the pipe wall by a rapidly revolving moving head inserted through the stationary pipe at the centerline. The lining is then cured by either storing it in a moist environment, processing it through an elevated temperature curing tunnel, or seal-coating it with a solvent-based paint. The lining produced by either of these methods produces a dense, smooth, surface that both inhibits interior corrosion and promotes flow. PDI is produced in accordance with ANSI/AWWA C104/A21.4. The thickness of the lining is to be not less than 1/16-inch (0.159 cm) for 3 to 12-inch (7.63 to 30.5), 3/32 (0.238 cm) for 14 to 24-inch (35.6 to 61.0), and 1/8-inch (0.318 cm) for 30 to 64-inch (76.2 to 163 cm) pipe. Normally acceptable operating temperatures are up to the boiling point of water unless seal coated in which case should not exceed 150°F (65.6°C). Sections of pipe are joined by bell and spigot with an intermediate rubber gasket placed in a groove inside the socket at the bell end.[3]

5.4.3 PLASTIC PIPES

The two main classes of plastic pipes are thermoplastics and thermosetting. Thermoplastic pipes soften at elevated temperatures above the useful range and become readily formable. On cooling, the material regains its original properties. Production by extrusion is simple, and joining and field modification exploit the softening and hardening ability. When heated to high enough temperature and pressed together, pipe ends fuse into a serviceable joint.[4]

HDPE pipe is a thermoplastic pipe manufactured by the extrusion process. Operating pressures can range from 50 lbf/in² (345 kPa) for large pipes to 200 lbf/in² (1,379 kPa) for small pipes. The operating temperature limit is 140°F (60.0°C). The pipes are manufactured according to ASTM D2104, D2239, D2447, F405, and F449.[2] Sizes covered by the standards are 0.75 through 200-inch (1.91. through 508 cm). Standard lengths range from 20 to 40 feet (6.1 to 12.2 m) with coils 1000 feet (305 m) in length.[2]

Among the thermoplastics, PVC pipe is the most common used for small plastic piping at power plants. It is also manufactured by the extrusion process. Operating pressures can range from 150 lbf/in² to 200 lbf/in² (1,034 to 1,379 kPa). The operating temperature limit is 140°F (60.0°C), but the strength is defined at 23°C (73.4°F) and decreases at higher temperatures. The pipes are manufactured according to ASTM D1785, D2241, D2321, D2412, D2672, and D3034.[2] (See also AWWA C900.) Sizes covered by the standards are 4 through 18-inch (10.2 through 45.7 cm).[2] Standard lengths range from 20 to 40 feet (6.1to 12.2 m) with coils 1000 feet (305 m) in length.[2]

Thermosetting resins cannot be reversibly cycled over a wide temperature range.[4] Thermosetting plastics require reinforcement, usually by fiberglass which forms a composite wall in which each layer contributes to the desired overall qualities of the pipe.[4] Thermosetting resin pipe reinforced with fiberglass reinforcement is manufactured by wrapping reinforcement fibers unbonded around the pipe surface and the pipe is then heated until the thermal bonding occurs.[4] Operating pressures can range from 30 lbf/in² (207 kPa) for large pipes to 250 lbf/in² (1,724 kPa) for small pipes.

The maximum operating temperature limit is 125°F (51.7°C).[2] The pipes are manufactured according to ASTM D2996 and ASTM 2997.[2] Sizes covered by the standards are 14 through 168-inch (35.6 through 427 cm). Standard lengths range from 40 to 60 feet (12.2 to 18.3 m).[2]

FRP pipe is a composite material made from a thermosetting polymer matrix reinforced with fibers.[4] Operating pressures can range from 30 lbf/in[2] (207 kPa) for large pipes to 250 lbf/in[2] (1,724 kPa) for small pipes.[2] The operating temperature limit is 140°F (60.0°C).[2] The pipes are manufactured according to ASTM D3517 and ASTM 3754.[2] Sizes covered by the standards are nominal 8 through 144-inch (20.3 through 366 cm).[2] Standard lengths range from 10 to 20 feet (3.05 to 6.10 m).[2] In some installations, the cost of the fittings to join the pipe runs may equal or exceed the cost of the pipe. Older installations of FRP pipe have experienced failures due to a lack of uniform soil compaction, manufacturing defects, and the use of unsuitable resins.

Considerable operating experience exists with plastic pipe in a variety of applications in industry. Plastic pipe is not generally subject to corrosion, but if the cooling water contains particles such as sand, erosion may occur mainly at turns. Plastic pipes are light and easy to handle, but extreme care must be taken in their installation in buried applications, as excess soil loading and/or ovalling of the pipe can be a problem.[5]

5.4.4   STAINLESS STEEL PIPES

Alternative pipe materials are normally less expensive than SS pipes, but may be used in buried applications. Although Type 316L has good resistance to chlorides, overall consideration of soil variability, halide leaching, etc. suggests that an exterior coating such as a heavy coat of halide-free paint or tape followed by a bituminous coating is advised.

5.4.5   IN SITU LINED PIPES

There are several types of in situ pipe lining processes. One is lining clean pipe with a thin coat of epoxy. A second is lining the pipe with a thick in situ tube saturated with epoxy that serves as a pipe within a pipe. A third is cement-mortar lining (CML). All three of these lining processes require access points in the buried piping some distance apart.

An epoxy coating alone lends no additional structural strength to the pipe. In order for epoxy coating to be successful, the pipe must be sand blasted to bright metal before applying the epoxy, a challenging task. Any failure to achieve bright metal or any pipe imperfection such as those brought about by rolling and finishing the pipe can result in imperfections in the lining process. Such imperfections may result in preferential galvanic attack at that point in the pipe and/or failure of the lining. Such coating failures frequently result in the epoxy coming off in sheets and blocking downstream valves and/or HX tubes.

In the second lining method shown in Figure 5.3, an engineered felt in situ tube or sock of the same diameter as the pipe and slightly longer than the distance between

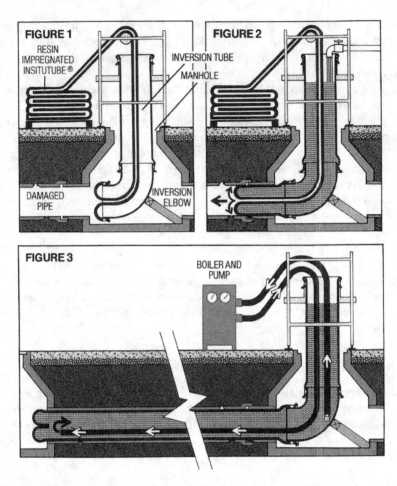

FIGURE 5.3   Resin-impregnated pipe lining. (Courtesy of Insituform).

access points is first impregnated with a liquid thermosetting epoxy resin and then folded. The open end of the in situ tube is attached to an inversion tube extending down from the surface to one end of the pipe. As the tube is filled with water, the in situ tube is inflated by the water, causing it to be turned inside out and pressing the resin-impregnated side of the in situ tube firmly against the inside wall of the pipe. The water inside the in situ tube is then circulated through a boiler, causing the resin to cure over a period of a few hours and changing the in situ tube into a hard, structurally sound pipe-within-a-pipe. After the water is pumped out and the inversion tube is removed, the ends of the in situ tube are cut off to match the ends of the pipe. This method of lining may be applied to pipe diameters from 6 to 60-inches (15.2 to 152 cm). The thickness of the lining is a function of the diameter of the pipe and depth of burial ranging from 0.1 to 1.45 inches (0.254 to 3.68 cm). No structural strength is required of the host pipe, since the resulting epoxy coated pipe has considerable strength, which is a function of the diameter to in situ tube thickness ratio.

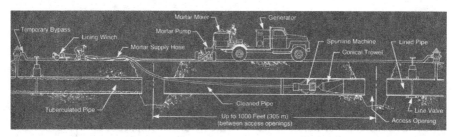

FIGURE 5.4   Cement-Mortar lining process. (Courtesy of Ameron Corp.)

The third lining method shown in Figure 5.4 is in situ CML.

A portion of the buried pipe is excavated and removed at intervals of several hundred feet. The spool piece thus removed is set aside and cleaned and lined by hand. The remaining buried pipe is scraped clean of corrosion products and other foreign material by pulling a "pig" through the pipe. Next, a lining machine is pulled through the pipe, applying a 0.25 to 0.5-inch (0.64 to 1.28 cm) layer of high strength mortar to the buried pipe, which is trawled smooth behind the lining machine in the process. The section of pipe is then sealed and cured in a high humidity environment for several days before the spool piece is welded back in place. The portion of the new lining damaged by the welding process is removed and the heat-effected zone is lined by hand. Obviously, if access to the interior of the pipe is required to repair the portion of the lining damaged by the weld, this process is principally employed in piping 24-inch (61.0 cm) and larger.

As discussed in Chapter 7, flow tests conducted at TVA's Sequoyah and Watts Bar nuclear plants in 1982 indicated that the large essential raw cooling water (ERCW) system headers between the IPS and the nuclear plants were exhibiting Hazen-Williams 'C' factors of from 75 to 85 after only approximately five years of service. These values were substantially below the C = 100 which was assumed in the original design. As Bowman and Guthrie[6] have shown, MIC, largely a pitting corrosion mechanism, can occur at a rate that is three times the general corrosion rate in cooling water piping, and the Tennessee Valley Authority (TVA) nuclear plants were observing MIC corrosion rates as high as 16 mils per year (MPY). This corrosion rate would imply a useful life of less than 23 years for large headers that are typically 0.375-inch (.9325 cm) thick. As part of the corrective action to address these deficiencies, TVA investigated the use of CML applied in situ, and this solution was eventually implemented at TVA's Watts Bar Nuclear Plant (WBNP).[7]

TVA was the first utility to apply a CML to a safety-related cooling water piping in situ. The project was undertaken only after exhaustive investigation. Factors considered in this investigation included operating experience with CML, the technical feasibility of the undertaking, the ability to qualify the lining, the ability to meet all codes and standards applicable to the ERCW system, the impact on the existing design, potential nuclear safety issues, the licensability of the option, the installed cost, and the impact on the plant schedule.

Miller[8] reported that CML has been employed successfully in municipal water systems for over 100 years. Field experience of the lining showed that CML was still effectively serving the purpose for which it was intended for many years. TVA surveyed the use of CML not only in municipal applications but also in fossil and nuclear power plants and discovered that CML had been successfully used in non-safety-related applications at numerous fossil and nuclear plants[7]. In general, the experience with in situ CML has been excellent when applied in accordance with the applicable AWWA standard.[9] However, experience with CML in salt water applications has not been encouraging. In situ CML should not be confused with pre-lined pipe which is welded or bolted together in the field. The impact of applying CML on the existing piping design is minimal, since the piping fittings and other access openings are reinstalled as originally designed.

The technical feasibility of CML is a function of the size or piping to be lined and the complexity of the system. Although pipe sizes as small as 4 inches (10.2 cm) have been lined, TVA decided that for safety-related applications, only those lines that can be accessed from inside the pipe should be candidates for CML. Since each valve and fitting must be removed from the system prior to lining, CML lends itself to long runs of large, buried piping.

The WBNP ERCW system is divided into two trains (A and B) with a WBNP Unit 1 train A and train B and a WBNP Unit 2 train A and train B supply header and a train A and train B discharge header serving both units. The portion of the WBNP ERCW system lined with CML consists of the four supply headers extending from the IPS to the plant auxiliary building and the two discharge headers from the auxiliary building to the cooling towers. The supply headers range in length from 5,600 to 5,900 feet (1,710 to 1,800 m) and in size from 24 to 36 inches (61.0 to 91.4 cm). The two discharge headers are 1,200 and 1,700 feet (366 to 518 m) long. The first 153 feet (46.6 m) of the supply header is 24-inch (61.0 cm). The balance is 30 and 36-inch (76.2 and 91.4 cm). Discharge headers are 36-inch (91.4 cm) in diameter. The project was divided into two halves with each half consisting of one of the supply headers serving each unit and one of the discharge headers in the same train being lined at a time so that a train of ERCW would be available to support construction at all times. An interval between lining the two halves was provided to complete the curing of the last section, closure, flushing, testing the first half, and unwatering, and opening the first section of the second half.

The CML process is a highly specialized construction technique that should be attempted only by those who are experienced in the practice. TVA allowed the CML contractor the freedom to use the same methods which had proven so successful in other applications. Every effort was made to minimize the impact on those techniques due to the fact that the project was safety-related.

The locations of access openings were selected to excavate valves and fittings and at high and low points in the piping and to ensure that no run of piping is longer than a few hundred feet. Where possible, locations were selected to minimize the depth of access holes. The excavations extended at least a foot below the pipe invert and exposed at least 12 feet (3.66 m) of pipe.

After the piping system had been unwatered, the valves, fittings, or spool pieces (which are referred to as closure pieces) were removed by cutting the pipe to permit access to the piping for cleaning and lining equipment. Spool pieces removed from straight runs of pipe were at least 8 feet (2.44 m) long. Each closure piece was removed from the hole and set on the ground beside the hole. Where fittings were removed, at least 2 feet (0.61 m) of pipe on the end or each run was exposed. After the piping was cleaned, lined, cured, and inspected, TVA welded the closure piece back into place. After the closure piece was welded into place and before the CML was applied by hand to the interior of the weld joint, a vacuum box test was performed on the weld.

After the access openings were created, the interior surface of the piping was cleaned by dragging spring-loaded mechanical scrapers through the pipe. A mortar mix consisting of Portland cement, Type II sand, and water was centrifugally applied by machine to each run of pipe so as to achieve a densely packed and uniform lining. The mortar mix was approximately one-to-one by volume, and the water/cement ratio was kept as low as possible. The mortar application was followed in the same pass by either a conical-shaped drag trawl or by rotating paddle trawls. Rotating paddle trawls were found to be superior to the conical-shaped drag trawl. With a drag trawl, the smallest piece of foreign material could be dragged through the pipe, requiring repair by hand. As the trawling machine moved along at a rate of approximately 4.5 ft/min (0.0229 m/s), the mortar was pumped to the machine through a hose in front of the machine. The travel speed, hose pressure, and trawl adjustments were set to achieve a 3/8-inch (0.952 cm) thick lining with a tolerance of +1/8-inch (0.318 cm) and -1/16-inch (0.159 cm). In areas where machine lining was not practical such as in fittings or closure weld joints, the lining was applied by hand. Curing was begun immediately upon completion of a lining run. The goal of curing was to maintain at least a 90% relative humidity inside the pipe. After the lining had been inspected and repaired as required, the closure piece was installed by welding. The heat-affected mortar was removed by chipping, and the weld joint was lined by hand from within the pipe. A curing compound was used only on this weld joint repair. Since CML does not rely upon its adhesion to the pipe but rather to the arching action to keep the CML in compression, its use in noncircular transitions is questionable without special design provisions.

TVA was responsible for opening and closing the holes and piping and keeping holes unwatered except when the CML contractor was in the hole performing cleaning lining activities. When the holes were backfilled, the same procedures were used as in the original construction. This included controlling the quality of backfill material and using vibrating compactors. Care was taken to ensure that the compactors did not ride against the pipe. The backfill activity was delayed until the lined pipe and closure piece had been cured for at least four days.

At points in the piping where the CML terminated, TVA installed termination rings consisting of 3/8-inch (0.9525 cm) thick by one-inch (2.54 cm) wide rolled bar stock that were fillet welded to the interior of the pipe before the CML was applied. In this way, the end of the CML butted against termination rings rather than being exposed to hydraulic forces. All ASME code activities were performed by TVA, since the CML contractor was not an "N" stamp holder. (An "N" stamp holder is authorized to perform work under ASME Section III at a nuclear facility.)

Proper curing is the most important step in the lining process. Curing was the responsibility of the CML contractor until a section of pipe had been inspected and accepted by TVA. Pipe end caps were used to keep the CML from drying too quickly.

After the lining achieved initial set, sandbags were installed at the ends of the pipe and water was added until it flowed steadily through the bottom or the pipe from one end to the other. The end caps were secured on the ends of the pipe for a minimum of four days with water being added daily in a similar manner. Any of the exterior surfaces exposed to the sun were covered with burlap that was wetted hourly. The closure pieces were capped, wrapped in plastic, and covered with wet burlap.

A total of 612, 13,083, and 12,238 feet (186.5, 3,988, and 3,730 meters) of 24, 30, and 36-inch (61.0, 76.2, and 91.4 cm) piping were lined. A total of 31 access holes were excavated to permit removal of 72 closure pieces. Approximately 200 cubic yards (153 m^3) of corrosion products were removed from the interior of the piping.

The project contractor cost in 1982 was $467,000. The cost of TVA support was estimated to be approximately $1,000,000 which included the cost of excavating and backfilling the access holes, cutting and welding the piping, curing the pipe, and performing inspections.

5.5 HUNG PIPING

Unlike buried piping which is bedded and supported by the earth, hung piping is simply supported by pipe hangers, and more rigid pipe are required to minimize the number of hangers. In hung piping, dynamic and thermal loading are concentrated at the hangers. If oversized CS piping is employed to compensate for the reduced flow-passing capability, the result can be lower velocities, resulting in increased sedimentation and corrosion. PDI pipe is generally not suitable for service in hung piping due to a lack of flexibility of the cement lining in these piping systems and the increased weight, especially at the joints where flanges are often employed. Similarly, CPP is not normally employed in hung piping for much the same reasons. Neither PDI nor CPP can be readily modified in the field, potentially resulting in schedule impacts due to field changes.

Dipped and heat-fused plasticized PVC coated pipe has successfully controlled corrosion in CS piping and fittings. Success depends on the proper procedures to ensure a good application of the coating. When properly applied, PVC coating has performed well due to its excellent adhesion characteristics. This tremendous adhesion strength results in a coating providing bond strengths that may exceed the tensile strength of the PVC itself. The coating does not delaminate or come off in large pieces as has sometimes been the case with in situ epoxy coatings. Before the coating is applied, the base metal is sand blasted to a near white metal finish and the coating is seamlessly baked on to the pipe making it impervious to attack by saltwater or chlorine. PVC coated pipe has had to be replaced in some areas subject to high cavitation. Because PVC is a thermoplastic, it will distort under high heat and pressure. If the lining is too thick on the flange face, it will possibly extrude into the inside diameter of the pipe if excessive torques are used during the assembly process. For this reason, the thickness on the flange face is kept as thin as possible. The PVC material has a tendency to stick to itself, making disassembly of spool pieces difficult.[10]

The use of CS piping clad with a corrosion-resistant alloy such as 300 series SS or Inconel 625 might be a suitable choice for replacing CS hung pipe.[11] Since the allowable stresses and thermal expansion would be essentially the same as the CS pipe, the material may be replaced like-for-like without requiring redesign and/or reanalysis.[11] For cooling water from rivers or lakes with hard water and low chloride levels, the 300 series SS would be adequate, while those from brackish or salt water may require a more corrosion-resistant material such as Inconel 625.[7] The cladded pipe is manufactured by a machine-gas arc welding and roll bonding process that produces a metallurgically bonded alloy 80 to 100 mm thick internal cladding.[7] The CS piping constitutes the pressure boundary, while the cladding constitutes the corrosion allowance. The root pass of field girth welds is made using the same alloy rod as the cladding, so the surface exposed to the cooling water after welding is the same as the cladding.[7]

Plasticized PVC coated pipe and alloy cladded pipe are normally flanged or screwed fittings, since welding would destroy the coating or cladding Flanged joints require more surrounding space for fitting up and tightening the bolts than do welded joints.[5]

In light of its resistance to corrosion, SS is the preferred material for new hung piping. Type 304 and 304L SS provide good service in many instances. In the presence of sedimentation, algae, and oxidizing biocides such as chlorine, Type 316L SS may be the preferred choice to deal with crevice corrosion and potential localized high concentrations of chloride ions leading to pitting. Brackish or salt water applications may require a more corrosion-resistant material such as Inconel 625.

Recently, plastic piping systems have been used inside power plants, but in addition to requiring more hangers, they present a fire hazard. Figure 5.5 shows HDPE supported by pipe hangers inside the Catawba Nuclear Station.

FIGURE 5.5   Cooling water HDPE supported by pipe hangers inside the Catawba Nuclear Station.

5.6   NUCLEAR SAFETY CONSIDERATIONS

5.6.1   SEISMIC QUALIFICATION

From a design standpoint, just about any buried pipe (steel, concrete, plastic, or lined) can be seismically qualified as long as the design stresses in the piping and joints are below the allowable stresses. Since seismic loads are mainly due to seismically imposed deformation, the more flexible pipes are easier to qualify, as flexibility reduces the loads in the pipe. Operating experience and tests on lined pipes indicate that most liners will adhere to the pipe and maintain their integrity while the parent pipe is deforming.

Although prior to the TVA project no effort had ever been made to seismically qualify CML, impressive evidence exists as to the ability of CML to withstand earthquake as long as there is no plastic deformation of the pipe base metal. During the 1971 San Fernando earthquakes, a 96-inch (244 cm) above-ground water line owned by the Los Angeles Department of Water and Power that was located within three miles of the quake epicenter suffered both vertical and horizontal displacement due to surface acceleration. CML was undamaged except where the pipe accordioned, where spalling did exist. A 20-foot (6.1 m) diameter steel tunnel liner owned by Metro Water District of Southern California and buried 50 feet (15.2 m) below ground was not damaged. The pipeline owned by the Metro Water District of Southern California, which serves the Jenson treatment plant separated by about 3 inches (7.6 cm) in two places and then was driven back together with one section inside the other, although the water treatment plant was largely destroyed by the earthquake. The CML was damaged only where the pipe separated.[12]

The evidence suggested that unless there is plastic deformation in the pipe, the CML will not be destroyed. Since safety-related cooling water must meet the requirements of ASME Section Class 3, the piping base metal must meet all Code requirements and no credit may be allowed for the CML. The CML is considered to be a coating.

To demonstrate that CML will withstand design basis seismic events, TVA implemented a full-scale testing program consisting of laboratory field tests and vibration measurements.[12] A total of 100 feet (30.5 m) of 30-inch (76.2 cm) diameter pipe, 20 feet (6.1 m) of 18-inch (45.7 cm) diameter pipe, and 90-degree elbow were lined in the field. The lining materials and procedures were the same as those to be used in the CML projects. Cement-mortar specimens were tested for compressive, tensile, and flexural strength, modulus of elasticity, and density. A 40-foot (12.2 m) length of 30-inch (76.2 cm) CML was installed in a trench and as it was being backfilled, it was subjected to a dynamic loading of 36,000 lbf (160.1 kN) at 28 hertz from a vibratory roller as shown in Figure 5.6.

The piping was then transported to the Singleton Materials & Engineering Laboratory (now known as SM&E). As the piping was transported the 100 miles (161 km) from the construction site where the lining was done to SM&E where the tests were conducted, two accelerometers were mounted on two of the 30-inch (76.2 cm) pipes to monitor vibration experienced by the pipes during the trip. Maximum acceleration experienced by the bottom pipe was 0.6g and that experienced by top pipe was 2.1g. Both values were higher than the 0.18g safe-shutdown

FIGURE 5.6 Cement-Mortar lining being subjected to dynamic loading.

earthquake acceleration used for the design or TVA nuclear plants. The recorded maximum peak-to-peak accelerations were 1.2g and 3.8g, respectively. Dominant frequencies ranged from 15 to 70 Hz. For most large earthquakes, the dominant frequencies are in the range of 0.5 to 10 Hz. Lower frequencies indicate a buried pipe would experience fewer cycles of vibration during a real earthquake. No cracks due to vibration were found in any of the CML after unloading, and it was concluded that the lining had experienced more severe vibrations than from any potential earthquakes in terms of magnitude and number of cycles.

After the piping arrived at the laboratory, a 30-ft (9.14 m) section of 30 inch (76.2 cm) pipe was subjected to bending, cyclic loading, and drop tests as seen in Figures 5.7 and 5.8.

FIGURE 5.7 Cement-Mortar lining being subjected to cyclic loading and bending test.

FIGURE 5.8   Cement-Mortar lining drop test.

FIGURE 5.9   Cement-Mortar lining elbow bending test.

Cracks appeared in the lining only after the applied stress exceeded the tensile strength of the mortar. No lining failure occurred.

2-foot (0.61 m) long sections of 30-inch (76.2 cm) pipe were subjected to cyclic loading, torsion, drop, and impact tests as seen in Figure 5.8.

Cracks appear in the pipe at 15 foot (4.57 m) drop height. A 5-foot (1.5 meter) length of 30-inch (76.2 cm) pipe was welded to each end of the 30-inch (76.2 cm) elbow and it was subjected to bending tests shown in Figure 5.9

The elbow was loaded to 50,000 lb (22,700 kg) without failure. When the elbow was loaded to 54,000 lb (24,500 kg), the lining failed, and the elbow was shortened by 5 inches, 1.5 inches permanently.

FIGURE 5.10   Cement-Mortar lining three-edge bearing test.

A three-edge-bearing test shown in Figure 5.10 demonstrated that the CML is flexible.

The CML failed only after the formation of plastic hinges in the steel. The lining underwent considerable cracking prior to separation and failure.

From these tests, it is concluded that the test loadings applied to the CML were much more severe and broad-ranged than the design seismic loadings. Therefore, the CML lining in buried SWS piping is seismically qualified when applied in accordance with approved QA procedures.[13]

5.6.2   CODES AND STANDARDS

Reference 14 establishes piping in safety-related cooling water systems in nuclear plants as "Quality Group C" and states that the requirements of this group are fulfilled if the system is designed to ASME Section III, Class 3 requirements.[14] Prestressed and pretensioned concrete pipes are not covered by ASME Section III but equivalency may be established under provisions of Reference 14. ASME Code Case N-155-2 addresses FRP. ASME Code Case No. N-755 addresses HDPE pipe. In 2008, the NRC approved the installation of HDPE to replace buried safety-related CS cooling water piping. HDPE was subsequently installed in the safety-related cooling water systems at the Catawba and Callaway nuclear plants.

Although the welding activities of the WBNP CML project were done to ASME Section III, Class 3 requirements, the Article ND-6000 requirements presented a challenge, since it was impractical to perform hydrostatic testing on each weld before the CML was applied to the interior of the pipe weld joint. Neither Section III nor Section XI specifically address the hydrostatic testing of lined pipe. The solution was to close out the Section III activities on the basis of the hydrostatic test that was performed prior to starting the CML activities and to declare the CML to be a "replacement" under Section XI and perform the test under Section XI after the CML has been applied. At least two code interpretations seemed to TVA to set a precedence permitting pressure tests in this manner, and concurrence was received from the Authorized Nuclear Inspector on this approach.

5.6.3   NRC APPROVAL OF CML

On April 30, 1982, TVA presented the WBNP CML project including the seismic qualification test results to the NRCs Advisory Committee on Reactor Safeguards (ACRS). The ACRS' primary concern was with the long-term durability or the CML with the potential for the failed lining to plug the tubes of HXs in the ERCW system. The ACRS recommended to the NRC staff that TVA develop in-service inspection requirements for the CML. Accordingly, the NRC requested additional justification for the modification. TVA responded with a report that addressed each of the following potential failure mechanisms: leaching, erosion, and spalling.

The TVA report pointed out that 9,200 lb/in^2 (63,400 kPa) cement mortar is extremely dense and relatively impermeable to mildly aggressive water. Tennessee River water has an average pH of 7.3, a chloride ion content of less than 10 ppm, and a sulphate ion content of less than 20 ppm. Other materials that might attack the mortar are in negligible quantities. TVA noted that there will possibly be some surface leaching, but it was pointed out that Miller[8] reported 1/16-inch (0.159 cm) thick lining was almost leached away after 29 years in water with pH of 6.6 and a total hardness of 12 ppm. TVA concluded that the CML will not lose more than 1/16 inch (0.159 cm) over the life of the plant.

Regarding the potential for the lining to fail due to erosion, TVA pointed out that the maximum water velocity in the headers is approximately 12 ft/sec (3.66 m/sec) and the suspended solids are less than 20 ppm. Therefore, neither cavitation nor abrasion is expected to occur. Factors which could possibly cause spalling are excessive shrinkage, corrosion behind the liner, or gross distortion of the pipe. It is noted that after the piping system is placed in service, the mortar will expand rather than shrink. The pH of the mortar will prevent corrosion behind the liner, and gross distortion of the piping will not occur since the piping is buried.

After a second meeting with the ACRS on August 13, 1982, the NRC issued a favorable Safety Evaluation Report (SER)[15] on the CML modification. The SER acknowledges the good experience with CML in city water distribution systems and in power plants. The SER states that the tests performed by TVA demonstrate the adherence and flexibility of the freshly installed CML, and the SER concludes that significant short-term failure of the lining is unlikely. The SER noted that as CML ages it normally gains in strength and that rates of leaching depend upon the hardness and acidity of the water. The NRC staff reviewed the program proposed by TVA to periodically inspect the condition and calcium content of the CML samples and found it to be acceptable. The SER states that the NRC concludes that it is unlikely that CML will be significantly weakened by leaching during the plant life and that the NRC staff concludes that CML in the ERCW piping was acceptable.

5.7   VALVES AND FITTINGS

Valve bodies and fittings are generally made of forged or cast steel, cast iron, or ductile iron. The connection to piping may be flanged, screwed, or welded. For pipe sizes under 3 inches (7.6 cm), forged steel socket welding fittings are often employed. Elbows may be fabricated from piping by bending the pipe. Complex sections of piping such as headers are normally fabricated in the shop using saddles to reinforce fabricated tees where required.

5.7.1 VALVES

A valve is a device used to prevent, limit, or permit the flow of cooling water through a portion of the system. Swing check valves as shown in Figure 5.11 prevent the reverse flow of the cooling water and are most often found at the cooling water pump discharge to prevent back flow through an adjacent idle pump. A disk is hinged at the top of the valve, and flow through the valve keeps the disk open, while reverse flow or gravity causes the disk to swing shut, shutting off flow.

Butterfly valves are the most common valves in a cooling water system because they are simple, relatively inexpensive, weigh less, and are easier to operate and maintain than many other types of valves. Butterfly valves use a circular disk to rotate through 90° to go from full closed to full open and may be stopped at any point in between to throttle the flow through the line. There is very little pressure drop when the butterfly valve is wide open. Most butterfly valves in cooling water system service have elastomer lining seats. They are controlled with either a hand wheel or are operator powered using either an electric motor or pneumatic actuator. The common body types are lug style in which the lugs are threaded on each side to receive bolts from adjacent flanges, wafer style where the body is clamped between two pipeline flanges, and flanged ends.

Gate valves are used for applications requiring either fully open operation (requiring minimum pressure drop) or fully closed operation (requiring a tight shut-off) and should not be used for throttling. When operating, the disk moves up or down on the end of a threaded valve stem, driven by an electric motor or pneumatic actuator. Gate valves are either parallel type in which a disk slides between two parallel seats or wedge-shaped, which has two inclined seats. When fully open, the disk is fully withdrawn, making it possible for a pipe cleaning pig to pass unobstructed. The gate

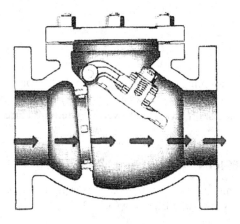

FIGURE 5.11   Swing check valve. (Courtesy Crane Co.)

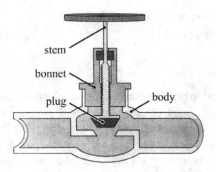

stem

bonnet

plug   body

FIGURE 5.12   Globe valve. (Courtesy Petteri Aimonen).

valve has either a metal or rubber seat in a depression in the valve bottom. Since for cooling water systems the potential exists for objects such as pebbles or mollusks to hinder complete closure of the valve, a rubber seat may be required if tight shut-off is required. Gate valves are more expensive than butterfly valves and require more physical space.

Globe valves as shown in Figure 5.12 are used to regulate flow by passing the flow through a spherical body in which the flow must pass through an internal baffled opening, the size of which is adjusted by a movable plug that can be screwed up or down. Typically, automated globe valves (control valves) use a smooth rather than threaded stem which is opened and closed by an actuator assembly that may be air-operated. Globe valves exhibit a relatively high pressure drop.

Ball valves are sometimes used in smaller lines where tight shut-off is impor-tant. A ball with a hole in it controls the flow. When the hole is aligned with the direction of flow, the pressure drop through the valve is insignificant, and when the hole is positioned perpendicular to the direction of flow, the flow is shut off. By positioning the ball in any intermediate position, the cooling water flow rate may be controlled by changing the pressure drop through the valve. Ball valves are nor-mally operated manually by a handle attached to a stem that controls the position of the ball.

5.7.2 FITTINGS

Figure 5.13 shows a sampling of the cooling water system pipe fittings that are com-mercially available. Long radius elbows are generally preferred because they require less pressure drop than short radius; however, they require more space.

5.8   PIPING DESIGN WALL THICKNESS

The required wall thickness of commercial grade piping under the ASME B 31.1 Power Piping code, is shown in Equation (5.1).[16]

90° elbows long radius, long tangent   45° elbows long radius, long tangent   saddles regular type   laterals reinforced and unreinforced   stub ends lap joint type   flange, ASA types welding neck, slip-on screwed, lap joint socket welding, blind

90° elbows short radius   180° returns long radius, ASA   shaped nipples 45-degree   reinforced rings for nozzles   sleeves butt-welding   large diameter welding neck, slip-on, blind

90° elbows long radius, reducing   extrusions headers and tees   crosses straight and reducing   caps   reducers concentric and eccentric   backing rings nub type

FIGURE 5.13   Pipe fittings. (Courtesy Crane Co.)

$$t_m = \frac{P\,D_o}{2\left(SE + Py\right)} + A \tag{5.1}$$

where

t_m = minimum required wall, in

P = internal pipe pressure, lbf/in^2

D_o = outside diameter, in

S = maximum allowable stress, lbf/in^2

E = joint efficiency (1.0 for seamless pipe)

y = temperature coefficient (0.4 for cooling water piping)

A = corrosion allowance, in (See Chapter 6).

Additional pipe wall thickness may be required to withstand external loads such as gravity, thermal loads, torsion, and seismic effects, etc.

The formulae for ASME Section III, Class 3 piping are the same as Equation (5.1).

REFERENCES

1. Linsley, R. K. et al. *Water-Resources Engineering*, 4th ed., p. 378, McGraw-Hill, Inc., New York, NY, 1992.
2. Chandley, C. A. et al. *Essential Raw Cooling Water (ERCW) System Piping Materials, memorandum MEB '810413022*, Tennessee Valley Authority, Knoxville, TN, 1981.

3. Bonds, R. W. *Corrosion Control – Cement-Mortar Lining for Ductile Iron Pipe*, Ductile Iron Pipe Research Association, Birmingham, AL, 2017.
4. O'Keefe, William. Corrosion-resistant Piping for Utility and Industrial Power Plants. *Power*, Vol. 125, No. 4, pp. S1–S24, April 1981.
5. Chandley, C. A. et al. *Nonsafety-Related Raw Water Systems Piping Materials, memorandum MEB '820107011*, Tennessee Valley Authority, Knoxville, TN, 1982.
6. Bowman, C. F. and P.V. Guthrie, Jr. *Corrosion in Carbon Steel Service Water Piping. 1994 ASME Pressure Vessel and Piping Conference*, Phoenix, AZ, ASME, 1994.
7. Bowman, C. F. *In Situ Cement-Mortar Lining of Safety-Related Service Water Piping Systems. International Joint Power Conference*, 94-JPGC-NE-6, 1994.
8. Miller, W. T. Durability of Cement-Mortar Linings in Cast-Iron Pipe. *Journal AWWA*. American Water Works Association, Denver, CO, June 1965.
9. AWWA Standard C602-76. *Cement –Mortar Lining of Water Piplines – Inch and Larger – In Place*, American Water Works Association, Denver, CO, 1976.
10. Robinson, L. K. and P. C. Fritchman. *Lining of Raw Water Piping with Plasticized PVC, EPRI Service Water Systems Reliability Improvement Seminar*, Electric Power Research Institute, New York, NY, 1994.
11. Chakravarti, Bhaven. *Use of Clad Piping Products for Solving Nuclear Plant Service Water Systems Corrosion, 91-JPGC-NE-12. International Power Generation Conference*, ASME, 1991.
12. Hubble, J. D. *Operating Experience for Alternative Raw Water Piping Materials, Memorandum MEB 810720032*, Tennessee Valley AuthorityKnoxville, TN, 1981.
13. Sun, C. N., et al. *Full Scale Testing and Qualification of Cement-Mortar Lined Carbon Steel Pipe*, CEB-82-8, Tennessee Valley Authority - Division of Engineering Design – Civil Engineering Branch, Knoxville, TN, 1982.
14. Regulatory Guide 1.26: Quality Group Classifications and Standards for Water, Steam, and Radioactive-Waste-Containing Components of Nuclear Power Plants, Rev. 4, Nuclear Regulatory Commission, Washington, DC, 2007.
15. Watts Bar Nuclear Plant SSER 4. *Report of the Advisory Committee on Reactor Safeguards*, Section 19, pp. 19–1 to 19–2, 1982.
16. ASME B31.1-2001. *Power Piping: ASME Code for Pressure Piping, B31.1 – American National Standard*, The American Society of Mechanical Engineers, New York, NY, 2001.

6 Corrosion and Fouling

6.1 CORROSION OF COOLING WATER PIPING

6.1.1 BACKGROUND

In the 1970s, a number of articles appearing primarily in the *Journal of the American Water Works Association* recognized the problem of tubercle deposits occurring in cast iron potable water distribution lines. It was recognized that these tubercles resulted from pitting corrosion, because they consisted largely of iron oxides and were found overlying pits. Past studies suggested that microorganisms accelerated the corrosion reactions and enhanced tuberculation. However, these early studies were primarily concerned with the problems associated with red water stemming from the oxidized iron released into potable water supplies during periods of hydraulic upsets and high flow demand. The iron content was at times sufficient to stain ceramics and clothing. Although screening showed no evidence for the presence of bacteria as a public health concern, an increase in heavy metals accompanied material loosened from the tubercle exterior during hydraulic stress. The potential was also recognized for esthetic degradation, including red water and taste problems. The loss of hydraulic capacity was also noted. The early publications did not focus on the impact that the corrosion was having on the structural integrity of the piping.

6.1.2 TENNESSEE VALLEY AUTHORITY INVESTIGATION

During preoperational testing of the Emergency Equipment Cooling Water (EECW) system at the Browns Ferry Nuclear Plant (BFNP) during the summer of 1976, certain HXs were found to be receiving inadequate cooling water flow due to a buildup of foreign materials on the interior of the CS piping servicing the equipment. A study was undertaken by TVA to determine the pervasiveness of this problem in the TVA system and to develop recommended practices to mitigate its effects in the design of future power plants.[1] A CS raw water piping sampling program was initiated to determine the extent of the problem in TVA power plants existing at that time. Approximately 50 piping samples were removed from nine different power plants and analyzed by the TVA Central Laboratories to determine the chemical composition of the buildup, the average pipe inside diameter reduction, the average pipe wall reduction, and the maximum pipe wall thinning. Although large differences were found in some of the above parameters between samples removed from the various plants, the problem was found to be widespread. In fact, difficulty was frequently encountered in locating original pipe samples in many of the plants because much of the CS raw water piping had become unserviceable and had been replaced.

One foot long (0.305 m) pipe samples were removed from not only CS cooling water lines but also fire protection lines, etc., in both flowing and stagnant service from existing TVA coal-fired power plants including Colbert (C), Gallatin (G), John

Sevier (JS), Watts Bar (WB), Widows Creek (WC), Kingston (K), and Cumberland (CU) and nuclear plants including BFNP and SNP. The sections of piping removed from the various raw water systems were kept hydrated and delivered for analysis to the TVA Power Service Center Laboratories in Chattanooga, Tennessee. A measured length of a piping sample was sealed at one end and filled with water. The volume of water contained in the sample was compared with the original volume as calculated from the nominal dimensions of a new pipe. The percent volume occupied by the deposit thus determined represented the average loss in pipe cross-sectional area and was related to an average decrease in pipe diameter. A 4-inch (10.2 cm) length of each pipe was split lengthwise, scraped, and acid cleaned in inhibited hydrochloric acid. The sections were weighed and compared to the weight calculated for new pipe of nominal size. The change in pipe weight was then directly related to the change in pipe wall thickness. A visual inspection of each piping sample was made after cleaning to identify the deepest pit or area of maximum wall thinning. The remaining wall thickness in the deepest pit was measured with a dial micrometer with a 1/32-inch (0.079 cm) diameter anvil. The deposit in each sample was analyzed for various constituents.

Figure 6.1 is illustrative of the amount of corrosion products and other material found inside of the piping. Although such pictures are very common in the technical literature today, when the piping was removed and examined by the author in the late 1970s the implications were staggering.

Observed and calculated data for the piping samples are shown in Table 6.1. The data in Table 6.1 is sorted by the number of years that the piping has been in service. Samples that had been in constant flow service are noted as "C" and those that had been in mostly stagnant flow service are noted as "S".

Figure 6.2 shows the average and maximum wall thinning from Table 6.1. As one may see, whereas the average wall thinning appeared not to be a strong function of age, the maximum wall thinning (i.e. pitting) was. The average ratio between the average and maximum wall reduction was approximately 3.

FIGURE 6.1   3-inch pipe removed from widows creek fossil unit 8.

TABLE 6.1
Analysis of Samples Taken from Deposits in Raw Water Piping

Sample ID	Flow Cond.	Age (Years)	Ave. Wall Red. (in)	Max. Wall Red. (in)	Ave. Corr. Rate (MPY)	Pitting Corr. Rate (MPY)	Pit/Ave Ratio	Fe_3O_3	SiO_2	S	Mn_3O_2
SNP-1	C	2.0	0.010	0.036	5.10	18.00	3.53	87.5		0.4	
CU-7	C	5.0	0.014	0.057	2.80	11.40	4.07	89.0	9.0	0.4	1.9
CU-6	C	5.0	0.025	0.019	5.00	3.80	0.76	81.2	17.0	0.4	2.7
CU-2	S	5.0	0.011	0.064	2.20	12.80	5.82	88.9	8.0	0.6	1.5
CU-8	C	5.0	0.023	0.055	4.60	11.00	2.39	78.0	14.0	0.3	1.5
CU-4	C	5.0	0.014	0.027	2.80	5.40	1.93	85.2	9.0	.05	1.4
CU-1	C	5.0	0.020	0.014	4.00	2.80	0.70	88.5	8.0	0.4	1.8
CU-5	C	5.0	0.021	0.075	4.20	15.00	3.57	74.0	21.0	0.6	2.0
WC-17	C	10.0	0.035	0.064	3.50	6.40	1.83	86.0	7.0	0.1	2.5
K-4	S	12.0	0.018	0.038	1.50	3.17	2.11	82.5	6.6		1.6
K-5	S	12.0	0.033	0.068	2.75	5.67	2.06	79.8	4.5		1.0
K-6	S	12.0	0.032	0.07	2.67	5.83	2.19	71.5	4.3		1.3
WC-4	C	12.1	0.014	0.046	1.12	3.80	3.38	90.4		0.5	
WC-23	C	13.8	0.019	0.095	1.38	6.91	5.00	81.0	8.1		2.4
WC-22	C	13.8	0.018	0.053	1.31	3.85	2.94	79.1	6.9		2.4
WC-21	C	13.8	0.035	0.067	2.55	4.87	1.91	65.0	20.1		3.7
WC-24	C	13.8	0.017	0.045	1.24	3.27	2.65	56.3	27.4		5.2
WC-19	C	17.0	0.039	0.063	2.29	3.71	1.62	70.0	10.0	0.1	3.5
WC-20	C	17.0	0.031	0.059	1.82	3.47	1.90	87.0	4.0	0.1	3.5
WC-16	S	17.0	0.021	0.062	1.24	3.65	2.95	90.0	6.0	0.1	3.0
WC-18	C	17.0	0.033	0.07	1.94	4.12	2.12	78.0	7.0	0.1	2.6
JS-2	S	17.7	0.020	0.061	1.11	3.45	3.10	75.8		0.5	2.2
G-1	C	17.8	0.022	0.077	1.26	4.33	3.44	88.1		0.8	
G-2	S	19.5	0.028	0.138	1.44	7.08	4.93	78.2	9.9	1.9	1.5
G-3	S	19.5	0.024	0.16	1.23	8.21	6.67	74.8	10.3	1.8	1.6
G-4	S	19.5	0.024	0.092	1.23	4.72	3.83	82.6	8.3	2.5	1.4
G-5	S	19.5	0.031	0.102	1.59	5.23	3.29	75.5	8.2	2.2	1.4
JS-1	C	21.5	0.016	0.079	0.72	3.67	5.10	3.7		0.1	72.4
C-1	C	22.2	0.027	0.073	1.20	3.29	2.74	82.9		0.3	
JS-6	S	23.0	0.045	0.086	1.96	3.74	1.91	89.5	7.0	1.1	1.4
K-8	S	23.0	0.007	0.042	0.30	1.83	6.00	79.4	4.8		1.1
JS-7	C	23.0	0.048	0.096	2.09	4.17	2.00	69.0	9.0	0.4	3.7
JS-8	C	23.0	0.028	0.04	1.22	1.74	1.43	44.0	13.0	0.3	4.2
K-7	S	23.0	0.017	0.143	0.74	6.22	8.41	81.8	5.4		1.3
JS-3	S	23.0	0.024	0.044	1.04	1.91	1.83				
JS-4	S	23.0	0.040	0.046	1.74	2.00	1.15	91.0	6.0	1.3	0.6
JS-5	S	23.0	0.061	0.096	2.65	4.17	1.57	85.6	6.0	0.7	1.3
K-10	S	23.0	0.025	0.128	1.09	5.57	5.12	90.3	4.2		0.9
K-9	S	23.0	0.022	0.044	0.96	1.91	2.00	80.2	3.4		0.8
WC-2	C	23.8	0.030	0.122	1.24	5.13	4.12	38.6		0.6	
WC-15	S	25.0	0.006	0	0.24	0.00	0.00	63.4	11.0	.4	3.2
WC-6	C	25.0	0.026	0.065	1.04	2.60	2.50	89.0	133.0	0.4	3.1
WC-13	C	25.0	0.022	0.036	0.88	1.44	1.64	77.9	7.0	0.3	2.4
WC-7	C	25.0	0.018	0.06	0.72	2.40	3.33	80.8	8.0	0.5	2.3
WC-8	C	25.0	0.016	0.129	0.64	5.16	8.06	72.3	15.0	0.8	2.8
WC-10	C	25.0	0.015	0.052	0.60	2.08	3.47	80.0	7.0	0.7	1.6
WC-11	C	25.0	0.054	0.095	2.16	3.80	1.76	72.5	7.0	0.4	1.9
WC-9	C	25.0	0.012	0.045	0.48	1.80	3.75	71.5	18.0	0.9	2.8
WC-12	C	25.2	0.027	0.046	1.07	1.83	1.70	77.8	5.0	0.4	1.5
WB-7	C	33.0	0.028	0.079	0.85	2.39	2.82	90.0	7.0	0.8	1.3

(Continued)

TABLE 6.1 *(Continued)*

Sample	Flow	Age	Ave. Wall Red.	Max. Wall Red.	Ave. Corr. Rate	Pitting Corr. Rate	Pit/Ave Ratio	% Deposit Analysis Fe₃O₃	SiO₂	S	Mn₃O₂
								Fe$_3$O$_3$	SiO$_2$	S	Mn$_3$O$_2$
WB-6	C	33.0	0.024	0.057	0.73	1.73	2.38	81.8	8.0	0.3	1.5
WB-2	C	34.1	0.035	0.1	1.02	2.93	2.88	84.4		1.0	
WB-3	C	35.0	0.016	0.057	0.46	1.63	3.56	93.7	4.0	0.3	1.0
WB-5	C	35.0	0.011	0.019	0.31	0.54	1.73	89.7	5.0	0.5	1.2
WB-4	C	35.0	0.018	0.13	0.51	3.71	7.22	92.5	5.0	0.4	1.2
Average							3.05	78.1	11.6	0.6	3.5

C = Constant flow.
S = Normally stagnant flow.

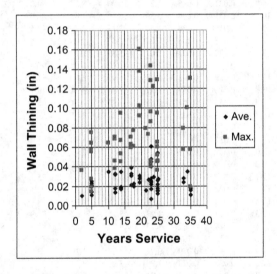

FIGURE 6.2   Pipe wall thinning.

From Table 6.1, it can be seen that JS-5, WC-1, JS-7, and JS-6 had the highest values of average wall reduction. Samples JS-5, JS-6, JS-7, and WC-11 all had varying degrees of exterior corrosion. This exterior corrosion also affects the maximum wall thinning calculations so that the maximum thinning for these samples would have been somewhat reduced if the exterior corrosion had been controlled. Varying degrees of exterior corrosion were noted on all these samples with JS-5 being the worst. Control of exterior corrosion would allow these values to be discarded. Data from stagnant lines were seen to fall within the same areas as the continuously flowing samples. Large differences in the average wall reduction are seen in samples removed from a given site and even between samples removed from a given pipeline.

Samples WC-15, WC-20, WB-8, WB-9, and WB-10 were removed from galvanized lines. Galvanizing appears to be more effective in controlling maximum wall thinning than in controlling average wall reduction. Much more data from galvanized

lines would be necessary before any conclusions could be made, however. The galvanized sample data should be used with caution since the maximum wall thinning
values are calculated from the remaining wall thickness measured by a micrometer.
The calculated values are underestimates of the actual thinning since the assumed
initial wall thickness of the pipe does not account for the additional wall thickness
due to the zinc coating. This results in negative values of maximum wall thinning for
some samples. WB-8 was found to be a unique galvanized sample. Removing the
insulation from the piping exterior to obtain the sample revealed a 0.4-inch (1.02 cm)
diameter hole rusted through the pipe wall. It is speculated that the zinc coating was
defective at this point, exposing the CS to the stagnant water in the line. The corrosive galvanic attack was accelerated due to the small area of CS pipe exposed to the
galvanized portion of the line.

As seen on Figure 6.2, large variations in maximum wall thinning were found
with the variations in data for samples from one site or from one pipeline at the same
site and even greater than that for the average wall thinning data. Values of maximum
wall thinning up to 0.160-inch (0.406 cm) were found in the observed data. The largest values of maximum wall thinning were seen in some of the 8-inch (20.3 cm)
samples at G (G-2 and G-3) and the 6-inch (15.2 cm) samples from K (K-7 and
K-10). Tubercles approaching 2 inches (5.1 cm) in height were found in these samples. It was found that areas of maximum wall thinning were usually found beneath
the large tubercles, indicating that maximum wall thinning is a function of tubercle
size. The average diameter reduction of these samples was less than some of the
smaller diameter samples due to only isolated instances of the large tubercles.

Figure 6.3 shows a plot of the average and pitting corrosion rate for each specimen
as determined by the method described above.

As one may see from Figure 6.3, both the average corrosion rate and the pitting
corrosion rate appear to decline with age, suggesting that cleaning the pipe without
adding an inhibitor may accelerate the rate of corrosion. The average corrosion rate
was 1.7 MPY.

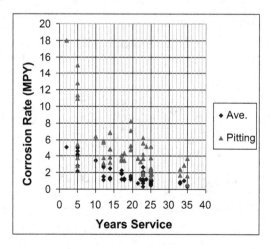

FIGURE 6.3   Average and pitting corrosion rate.

The scope of the 1979 TVA study was not intended to predict a 40-year design value of average wall reduction or maximum wall thinning (i.e. corrosion allowance "A" in Equation 5.1) or the impact on system design of the reduced wall thickness but only to report the observed data and to make any comments regarding trends or peculiarities. It was observed that the samples having large values of average wall reduction also had varying degrees of exterior corrosion. If all samples are considered, average wall reductions reaching 0.0625 inches (0.159 cm) were seen. Discarding samples with obvious extreme exterior corrosion drops the maximum value of the average wall reduction to 0.040 inches (0.102 cm). Values of maximum wall thinning up to 0.160 inches (0.406 cm) were found in the observed data. It is noted that only one sample was found with through-wall leaks, since they would normally have already been replaced. Therefore, the maximum wall thinning could be much higher than indicated by the 1979 study. No significant differences were observed in the corroded condition of horizontal versus vertical runs of pipes as long as the pipes were completely full of raw water.

In 1980, some of the results of the TVA investigation focusing on the flow-passing capability of the piping were reported by Bowman and Bain.[2] (See Chapter 7.) Figure 6.4 from Reference 2 shows the effective decrease in diameter of the pipes due to material buildup as determined by the method described above as a function of years of service.

One may see that the degree of material buildup varied greatly from sample to sample, but generally it appeared to be progressive with the age of the pipe. In stagnant, continuously pressurized piping systems, the rate of material accumulation is generally lower than that found in raw water piping systems where water flows in a continuous or nearly continuous manner, thereby replenishing the oxygen supply which can then induce further corrosion. However, data from some stagnant lines fall within the same areas as the data from flowing lines.

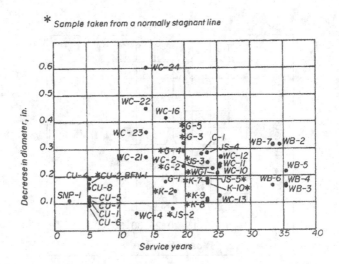

FIGURE 6.4 Average reduction in pipe diameter vs. years of service.

From the scatter of data seen on Figure 6.4, it can be seen that age was not the only parameter influencing corrosion product buildup. Large variations in buildup are seen for piping removed from a given site at a given age (e.g. WC and WB) and in some cases large variations can be seen from samples removed from a single pipeline (e.g. WC-21 through WC-24 and G-2 through G-5). It was found that the average buildup in 8-inch (20.3 cm) diameter piping (G-2 through G-5) and 6-inch (15.2 cm) diameter piping (K-7 through K-10) was on the same order of magnitude as the buildup in the 2, 3, and 4-inch (5.1, 7.6, and 10.2 cm) lines taken at JS and WC. Buildup did not appear to be dependent on pipe diameter.

Large differences in the appearance and consistency of the corrosion product buildup were found. In some cases, materials other than corrosion products were found on the pipe interior. At JS, some of the samples were found to have a large amount of manganese deposit. Sample WC-24 was found to have a higher level of silica than other samples. Most of the samples had a relatively uniform buildup with a very rough surface like that shown in Figure 6.1. However, some samples such as K-10 had almost no average buildup but had large, randomly spaced isolated tubercles. The samples removed from WB appeared to not be any worse than those removed from WC and JS even though the WB piping had been installed approximately 20 years earlier. However, It should be noted that WB was out of service for a significant period of time (10–15 years) and the status of the piping during that period is not known. The effective age of the piping at WB could actually be less than that shown. It can be seen, in most cases, that the diameter reduction in galvanized lines is less than that of the other samples. It can be seen from Figure 6.2 that age effect is only one part of the explanation for material buildup and that the worst buildup observed actually occurred between 20 and 25 years of service.

Reference 1 concluded that although the appearance and consistency of the buildup on the inside wall of the pipe samples varied, iron oxide was virtually always the principle constituent, and the primary mechanism was always corrosion of the steel piping by the aerated raw water with redeposition of the corrosion products onto the inside wall of the pipe in the form of irregular tubercles. The entire process was determined to be induced or at least influenced by sulfur-reducing bacteria (SRB). Corrosion of raw water piping and the resultant redeposit of ions of corrosion product onto the inside of the pipe were found to a significant degree at all plants that were sampled. Under each tubercle was almost invariably found an area of reduced pipe wall thickness. The results of the study indicated that at the end of the 40-year life of a plant in the TVA system, CS raw water piping will experience an average reduction in the inside diameter of 0.40 inch (1.02 cm), an average reduction in the pipe wall thickness of 0.065 inch (0.165 cm), and a maximum wall thinning of at least 0.160 inch (0.406 cm) due to corrosion. As a result of the 1979 TVA study (Reference 1) TVA issued References 3 by J. R. Alley as guidance for design engineers in dealing with CS piping in raw water service.

Reference 1 also presented the results of pressure drop tests performed in straight sections of CS raw water piping at three TVA power plants. These tests conducted to determine the impact of corrosion products buildup on the interior pipe wall on the flow-passing capability of the pipe are addressed in Reference 2 and in Chapter 7.

6.1.3 INDUSTRY RESPONSE TO MIC

TVA immediately notified the NRC of the flow deficiencies discovered at the BFNP and the results of subsequent investigations at SNP and WBNP through nonconformance reports, as noted in Chapter 1. TVA also authorized the publication of Reference 2 in 1980, which presented the results on Table 6.1 above but focused mainly on the problem of flow deficiencies. In 1981, TVA provided the NRC with a copy of Reference 1 along with their final report on the nonconformance reported in 1980 (NCR No. SQNNEB8035). The NRC did not formally notify the nuclear power industry of the MIC problem until 1985 with the issuance of Reference 4 which only alerted licensees of the potential for through-wall leaks in butt-welded SS piping in raw water service. The basis for this IN was a condition reported by Carolina Power and Light (CP&L) at their Robinson Nuclear Plant in which minor pinhole leaks were found in the heat affected zones of circumferential welds jointing 6-inch (15.2 cm) diameter, schedule 10, 304 SS piping that provides service water to the four containment chilling units. Subsequent visual inspection of the system revealed leakage at a total of 54 weld joints. CP&L determined that the root cause of the problem was the result of MIC. The NRC noted to the industry that the CP&L experience was only one of a number of similar incidents reported to them. The NRC did not warn the nuclear power industry of the MIC problem in CS cooling water piping until 1989 with the publication of Reference 5.

Although INPO issued Reference 6 in 1984, the first significant response to the call for a response to the problem of MIC was Reference 7 by G. Licina, sponsored by EPRI. Reference 7 provides a discussion of MIC along with 15 case histories. For each case investigated, Reference 7 reported the operating history, the method of discovery, and a characterization of the material in the pipe. The water chemistry was reported for some cases. The pictures of the piping samples are similar to Figure 6.1 and many others presently found in the literature on MIC, including some of MIC attacks on both CS and SS welds. Licina noted that in CS pipe, MIC may result in both general corrosion and random pitting, causing the formation of tubercles anywhere around the circumference of the pipe and on horizontal or vertical surfaces.

In 1993, EPRI completed their *Service Water System Corrosion and Deposition Sourcebook*, a comprehensive reference document. EPRI continues to support their members through additional publications and through regular meetings of their Service Water Assistance Program (SWAP).

In 1980, TVA undertook a significant program of material upgrade to address the corrosion problem. In 1982, the first through-wall pit in a cooling water system was discovered in a 3-inch (7.6 cm) CS pipe in the BFNP EECW system. The report concluded that the apparent cause of the pitting was the buildup of biological material attributed in part to the presence of Asiatic clams in the system. Multiple leaks were also discovered in the fire protection system at BFNP that year. The TVA began a program of ultrasonic testing (UT) inspection of 11 EECW system CS piping areas at BFNP. In 1989, UT inspections conducted on the BFNP CS fire protection system piping indicated unacceptably high corrosion rates.

In 1986 in response to Reference 4, a leak was found in a SS butt weld in the stagnant auxiliary feedwater supply line of the WBNP ERCW system. A metallurgical evaluation showed that a condition similar to that reported to CP&L existed. As a result of the discovery in the ERCW system at WBNP, inspections at BFNP and SNP revealed similar leaks in applicable corresponding systems at those plants. The radiographic testing (RT) of SS piping in the EECW system at BFNP indicated only minor MIC damage.

Figure 6.5 shows a 6-inch (15.2 cm) SS butt weld from SNP sectioned to show MIC damage. Nondestructive examination (NDE) of the damage to SS butt welds in the ERCW system at SNP suggested that, unlike MIC damage in CS resulting in relatively benign bowl-shaped pits, MIC pits in SS can lead to subsurface cavities in the butt weld and heat-affected zone that threaten the structural integrity of the piping. In addition, the large number of through-wall pits implied a very aggressive attack. When welds were removed and replaced, in some cases leaks appeared in the new welds within six months.

In SS piping, the MIC damage may be characterized as pits oriented in the locale of the butt welds. As may be seen in Figure 6.5, the pits are quite insidious because they exhibit very small entry holes with much larger subsurface cavities which frequently proceed through-wall with a small exit hole. The resulting leaks are on the order of a few drops per minute. The pits occur in random fashion around the circumference of the weld. The corrective action developed by Pavinich and Deardorff included the development of a methodology to assess the damage existing in the butt welds of SS piping to assure that the structural integrity of the piping was not violated before the damaged pipe was identified and replaced.[8,9,10,11]

Austenitic SS exhibits significant toughness properties, resistance to fatigue damage, and resistance to ductile tearing. Therefore, the failure mechanism assumed for gross structural failure under seismic loading is plastic collapse under limit load conditions. ASME Section III design margins against limit load plastic collapse may

FIGURE 6.5 Stainless steel butt weld showing MIC damage.

be calculated. The approach was an appropriate adaptation for raw water piping of the procedures found in ASME Section XI (IWB-3640) for evaluating flaws in austenitic SS piping in ASME Section III, Class I systems, which provides convenient but overly conservative means of evaluating MIC damage. In this approach, all MIC indications must be added together in the circumferential direction and evaluated as a single flaw oriented in the direction of the primary bending to produce the weakest net sectional properties. Additional refinements have been developed to eliminate this conservatism as required on a case-by-case basis. The approach taken is to establish a screening criteria of allowable total length of MIC defects based on the IWB-3640 methodology for a given line size and enveloping load magnitudes for a given system. All welds with a total MIC damage indication of less than this value are structurally adequate. Welds with greater MIC damage were evaluated individually, first by using the methodology with location-specific loads and then with the less conservative methods if required. It should be noted that all methods of evaluation assumed that all MIC indications were completely through-wall and make no allowance for sound metal that may exist above the MIC cavities. This approach was required because existing methods of NDE did not permit reliable evaluation of MIC damage in the radial direction at the time the flaws were evaluated.

In 1987, TVA began investigating corrosion rates by means of corrosion coupons. Although this technique cannot predict the long-term corrosion rate, it is useful to determine the relative short-term corrosion rate as a function of time, temperature, and various water treatments. TVA conducted tests at the SM&E Laboratory during which CS coupons were exposed in a once-through service water flow loop drawing water from the Tennessee River at Knoxville, Tennessee. The results of these tests are shown in Table 6.2.

These data are shown in Figure 6.6.

Blackburn and Mullin[12] reported on the corrosion rates measured in a condenser waterbox at the BFNP in 1987–1989. Figure 6.7 shows the corrosion rates measured at BFNP by an array of different durations of coupon exposure. As with Figures 6.3 and 6.6, the same trend of lower corrosion rates with time is observed. Although the magnitude of corrosion as a function of duration of coupon exposure rates varies considerably between the three studies, the data consistently suggest that the average corrosion rate is a function of service life and that the rate decreases with time.

TABLE 6.2
Coupon Corrosion Tests in Cooling Water

Start	End	Time	Corrosion Rate
Date	Date	(Days)	(MPY)
10/22/1987	11/23/1987	32	5.35
11/24/1987	1/4/1988	41	3.60
1/8/1988	2/12/1988	35	5.94
11/24/1987	3/21/1988	120	2.78
2/24/1988	4/7/1990	773	1.59
11/22/1987	4/7/1990	898	1.86

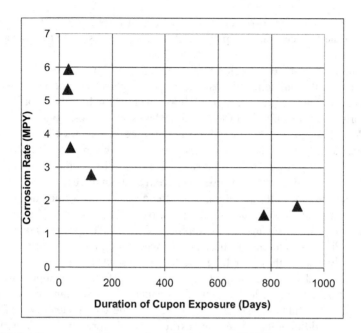

FIGURE 6.6 Coupon corrosion tests in cooling water at SM&E.

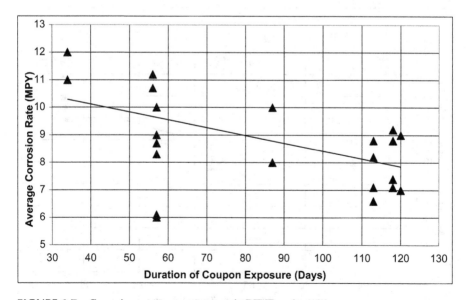

FIGURE 6.7 Corrosion rate vs. temperature in BFNP main condenser.

Figure 6.8 shows the corrosion rates measured at BFNP by 60-day coupons as a function of the service water temperature. This study suggests that the corrosion rate increases with increased cooling water temperature.

Although the average corrosion rates reported above are useful in assessing the various factors that influence corrosion, the pitting corrosion rate is the parameter that is of more interest to the designer. Unfortunately, much of the published literature reports on corrosion monitoring performed with weight loss coupons which normally do not reflect pitting corrosion rates. Figure 6.3 shows the corrosion rate calculated based on the deepest pit observed in the piping samples removed from the nine different TVA power plants as discussed above. When compared with average corrosion rates, pitting corrosion appears to occur at about three times the average for the same specimens. Although the general trend of corrosion rates declining with service life appears to hold true for pitting corrosion, caution should be exercised in drawing conclusions regarding an appropriate pitting corrosion rate based on Figure 6.3. Piping specimens that have experienced through-wall pits are not included in this sampling, since they have long since been replaced. Through-wall leaks began to occur in the EECW and fire protection piping at BFNP after approximately 13 years of service. The first leaks occurred in 3-inch (7.6 cm) standard wall piping, which implies a pitting corrosion rate of 16.6 MPY. These initial leaks were followed by 80 to 100 additional leaks over the next few years before several hundred feet of piping were replaced in 1988. In 1993, a through-wall leak occurred in a 16-inch (40.6 cm) standard weight pipe at BFNP, implying a pitting corrosion rate of 16.3 MPY. WBNP reported leaks in small-bore [2 to 6-inch (5.1 to 15.2 cm)] standard weight piping after about 14–17 years of service resulting in calculated pitting

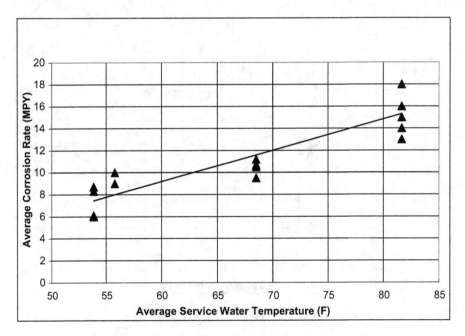

FIGURE 6.8   Corrosion rate vs. temperature in BFNP main condenser.

corrosion rates of 8 to 16 MPY. In 2001, several leaks were discovered in the 12-inch (30.5 cm) standard wall [0.375-inch (0.9525 cm)] WBNP fire protection supply header from the IPS after 23 years of service for a pitting corrosion rate of 16.3 MPY. The entire header was subsequently replaced.

In 1993, TVA initiated a comprehensive assessment, repair, and replacement of the raw water piping systems at the SNP. The results of the RT data for the SS butt welds showed a large percentage containing small, randomly distributed MIC defects. None of the 55 butt welds evaluated failed to meet the structural integrity criteria. The SS butt welds inspected in the ERCW system that showed indications of MIC were either monitored or replaced with Inconel 625 filler metal. Figure 6.9 shows the location of the only socket weld inspected that showed possible indications of MIC. An analysis of this weld at SM&E concluded that socket welds are less prone to MIC due to the location of the fillet weld. Since full penetration does not normally occur, the weld is not in contact with the cooling water.

Visual inspection of the interior of the SS butt welds where MIC damage had occurred showed that a small tubercle existed at the interior of the pipe at the weld joint. A water treatment program to eliminate or minimize the extent of attack at the SS butt welds by preventing the formation of these tubercles is discussed in Section 6.4.1.

Figure 6.10 shows the typical condition of the CS piping encountered at SNP. Random leaks have occurred in low flow areas of 2-inch (5.1 cm) and smaller cooling water and fire protection piping, and in some cases, when the corrosion products were removed, pits were found to be through-wall. CS piping was inspected by UT. The "as found" minimum pipe wall thickness was compared with the minimum wall thickness required to provide structural integrity. If the thickness was less than the required value, a specific minimum wall thickness calculation using point-specific stresses was performed. The UT data indicated random pitting throughout the CS systems. The more severe pits that were detected were small in diameter and clearly

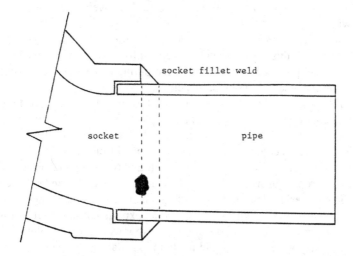

FIGURE 6.9   Socket weld with MIC damage at SNP.

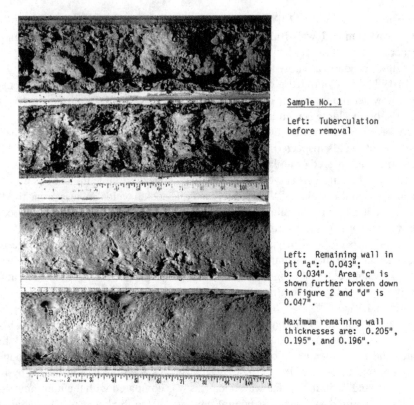

Sample No. 1

Left: Tuberculation
before removal

Left: Remaining wall in
pit "a": 0.043";
b: 0.034". Area "c" is
shown further broken down
in Figure 2 and "d" is
0.047".

Maximum remaining wall
thicknesses are: 0.205",
0.195", and 0.196".

FIGURE 6.10   Three-inch carbon steel RCW pipe from sequoyah nuclear plant.

isolated with considerably thicker base metal surrounding each pit. The study concluded that gross degradation leading to structural failure of CS piping is unlikely, but random leaks could reasonably be expected. The average corrosion rate in the cooling water systems at SNP was determined to be 4 MPY after approximately 17 years of service. However, some locations exhibited corrosion rates as high as 13 MPY. Although this effort provided useful information on the condition of the system, it did not preclude the possibility that pitting corrosion at even higher rates were occurring at other points in the system.

In 1994, Bowman and Guthrie[13] summarized the results of the TVA studies of CS corrosion in cooling water systems and the implications for piping system designers in selecting an appropriate corrosion allowance "A" in Equation 5.1 in the future.

Tuovinen, et al,[14] recognized that tubercles resulted from pitting corrosion because they consisted largely of iron oxides and were found overlying pits. Past studies suggested that microorganisms accelerated the corrosion reactions and enhanced tuberculation. The bacteria found in most cooling water systems make them particularly vulnerable to corrosion. Touvinen conducted an experiment in which cast iron coupons were immersed in various test solutions for four weeks. These solutions included tap water, distilled water, and a liquid medium inoculated with samples of red water and ground tubercles. Although coupon surface reactions took place under all assay conditions, the corrosion was twice as great in the inoculated medium. This led Touvinen to conclude that bacteria could be involved in the mechanism of corrosion.

There was some hope that most of the iron showing up in the tubercles had been scavenged from soluble iron found in the raw water, as was the case with the silica and manganese found in the deposits. However, that hope was refuted by a study conducted by Vermont Yankee Nuclear Power Corporation and published by Metell[15] in 1986. In that study, the author conducted tests on removable spool pieces of CS pipe over a five-year period. The masses of the pipe spools were determined before and after the test. The deposits were removed and the pipe was weighed once again to yield the mass of the pipe after the test and the mass of the deposits. A volume/density test and molar/mass balance showed that the deposits were composed of 10% base metal, 90% oxygen and water on a volume basis. A comparison was made between the oxide deposit based on the base metal loss and the actual oxide deposit. The good agreement between these values supports the conclusion of the study that the oxide was almost exclusively from corrosion and not deposition of soluble iron from the source stream. Further, the study observed that the deepest pits had the largest tubercles, that the iron lost from the pipe wall ended up in the overlying tubercle, and that the formed tubercle is approximately ten times larger than the deepest pit.. Metell observed that mechanical cleaning to remove tubercles to improve system flow characteristics promotes corrosion of the pipe wall.

In 1986, the American Nuclear Society published Reference 16 as an American National Standard for nuclear safety cooling water systems in light water reactors. The Standard stated that these systems should be constructed of corrosion-resistant materials, be protected by corrosion inhibitors or cathodic protection, or have sufficient wall thickness allowance to ensure the integrity of the system over the design life of the plant. Considering the fact that by that time, construction of virtually all of the nuclear plants in the United States had been completed, this document, though useful for future plants, was rather like closing the barn door after the horse had escaped.

In 1990, the NRC published Reference 17. For Code Class 3 piping, a licensee is required to perform code repairs or request NRC to grant relief for temporary non-code repairs on a case-by-case basis regardless of pipe size. Because of the rather frequent instances of small leaks in some Class 3 systems such as cooling water systems, the NRC provided guidance that would be considered by the NRC staff in evaluating relief requests for temporary non-code repairs of code Class 3 piping. The guidance consisted of assessing the structural integrity of the flawed piping by a flaw evaluation and assessing the overall degradation of the system by an augmented inspection. The document concluded that "temporary non-code repair of Class 3 piping that cannot be isolated without a plant shutdown is justified in some instances. The rather frequent instances of small leaks in some Class 3 systems, such as service water systems, could lead to an excessive number of plant start-up and shutdown cycles with undue and unnecessary stress on facility systems and components if the facilities were to perform a code repair when the leakage is identified."[17]

In 1991, the ASME published Reference 18 as a guide for determining the remaining strength for corroded pipelines (not power piping, e.g. B31.1). Unlike MIC in butt-welded SS where the damage is preferential to the transverse axis of the pipe (i.e. the weld) making the pipe more vulnerable to seismic and dead weight loads, the MIC damage in CS is random and catastrophic failures in the longitudinal direction may be possible. Although Reference 18 is not applicable to power piping, per se, the document contains guidance as to how to determine the maximum allowable longitudinal extent of corrosion.

In 2007, INPO issued a Significant Event Report on inadequate design resulting in cooling system leaks at several nuclear plants.

6.1.4　MIC Tubercles

In addition to the role that microorganisms play in corrosion of CS service water, general corrosion and the differential aeration cells set up by the tubercles are also factors.

General corrosion, present when CS is exposed to an oxidizing agent such as aerated water, can increase when an oxidizing biocide such as chlorine is used to control bacteria. (See Chapter 12.)[1] This requires the designer to balance the benefits that the biocide may provide in controlling the pitting corrosion from the MIC against the increase in general corrosion.

Although differential aeration cells can be associated with any form of crevice, in raw water they are most commonly associated with tubercles. Figure 6.11 shows a typical tubercle in CS pipe.[19]

All tubercles have five structural features in common:

1. Outer crust
2. Inner shell
3. Core material
4. Fluid-filled cavity
5. Corroded floor.[20]

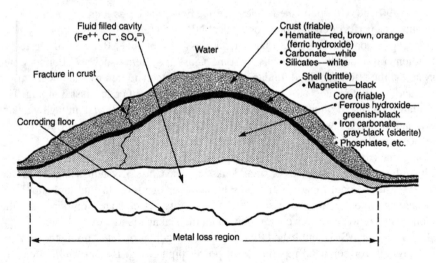

FIGURE 6.11　Tubercle in carbon steel piping.

As discussed by Metell[15] and Herro,[19] much of the tubercle material is formed of corrosion products from the underlying metal. The tubercle often contains an inner cavity filled with fluid. As tubercles increase in size with age, free oxygen within decreases and differential aeration cell activity is promoted. Anions concentrate within the tubercle, lowering the pH of the fluid inside the tubercle even if the bulk fluid pH is alkaline. Since both differential aeration cells and MIC cause pitting corrosion, often occurring at the same site, distinguishing between the two in service water is often difficult. MIC is considered by many experts to be the most serious form of corrosion in service water systems. In the case of MIC, several types of bacteria often work together to attack the pipe. Slime-forming bacteria (psudemonas) produce a biofilm which becomes the host for iron-reducing bacteria, which are aerobic. This iron-reducing bacteria then can provide a covering for SRB, which are anaerobic. SRBs are often found inside tubercles and produce the characteristic odor of hydrogen sulfide when the tubercle is penetrated while still wet. This discussion is an oversimplification of the subject. All or none of the above may occur in a specific case.

6.1.5 POTENTIAL IMPROVEMENTS TO COMBAT CORROSION IN COOLING WATER SYSTEMS

The designer must be concerned with the impact that corrosion will have on the structural integrity and serviceability of CS cooling water piping. Since ASME Section III is a design code, it does not provide explicit rules for evaluation or acceptance of degradation due to corrosion. However, the design should retain the structural margins inherent to the design requirements stipulated in Section III by defining acceptable degradation due to anticipated corrosion. Subsection ND-3613.1 states "When corrosion or erosion is expected, the wall thickness of piping shall be increased over that required by other design requirements". In determining the minimum thickness of pipe wall required for the design pressure, Equation 3 of ND-3641 requires the designer to specify an additional thickness "A", which is to include an appropriate corrosion allowance to consider all forms of corrosion including pitting corrosion. (See Equation 5.1.) When considering sustained (deadweight, thermal expansion, etc.) and occasional (seismic, etc.) loads, Code pipe stress equations in Subsection ND-3650 stipulate the nominal wall thickness without any consideration for corrosion. For typical piping systems under normal or service level "A" plant conditions, axial stress usually does not control structural design margins. However, for service water piping designs with either a large corrosion allowance and/or a conservative size applied, moment loading may dominate the net required wall thickness, particularly for occasional loads. Long-term maintenance of design structural margins can be assured through in-service monitoring.

Figures 6.6 and 6.7 suggest that as the interior surfaces of CS pipes become coated with oxides, the average corrosion rate steadily declines with exposure time. Figure 6.8 suggests that the rate of MIC corrosion is influenced by the temperature of the cooling water as well as water quality and flow velocity. The results of any corrosion monitoring program must be interpreted in light of these factors. When considering an additional thickness to be applied to the pipe wall required for the design pressure, pitting corrosion should be considered. The pitting corrosion rate can be as much or more than three times the average corrosion rate. All three TVA nuclear plants have experienced through-wall leaks in CS piping. Based on the pipe wall thicknesses and service life of piping, a pitting corrosion rate of 13–16 MPY or more for untreated raw water is not unreasonable. Data obtained from an effective monitoring program can be beneficial in making reasonable predictions as to pipe integrity and operability at a given site. (See Section 6.4.)

The designer should seek to minimize the impact of MIC in designing the service water system. The system operating water velocity is one parameter that the designer should consider to help reduce the impacts of corrosion. Metell[15] reported that fast moving streams appear to interrupt tubercle formation, and that stagnant and low velocity water appears to deprive the bacteria of required oxygen, reducing growth. He stated that moderate velocities [2 to 7 ft/sec (0.61 to 2.1 m/s)] encourage bacteria growth. Figures 6.12 and 6.13 show 24-inch (61.0 cm) headers in the SNP ERCW system. In Figure 6.12, the normal cooling water velocity exceeds 10 ft/sec (3.05 m/s), whereas the flow in the redundant header shown in Figure 6.13 is normally zero. Exceedingly high normal velocities (over 10 ft/sec) (3.05 m/s) can lead to erosion in CS cooling water piping.

Normally, the sedimentation that would be associated with velocities below 3 ft/sec (0.91 m/s) would promote the growth of the anaerobic SRB and so most experts recommend that normally flowing systems be designed for at least 3 ft/sec (0.91 m/s) to keep solids in suspension. Provisions for chemical treatment such as injection points, special provisions to make sure that the chemicals will reach all parts of the system, and provisions for side-stream monitoring or spool pieces should be included in the system design.

FIGURE 6.12   24-inch cooling water header where velocity exceeds 10 ft/sec.

FIGURE 6.13   24-inch cooling water header with normally zero velocity.

Depending on the desired service life, standard wall pipe thicknesses may not be adequate for CS piping. This is especially true for piping 3-inch (7.6 cm) and smaller. Socket welded SS pipe may be employed that is inherently much less susceptible to MIC. If SS piping is used for larger piping requiring butt welds, an effective biocide treatment designed to prevent the formation of MIC nodules over the welds is imperative. (See Section 6.3.1.) For larger piping, CS may be acceptable for hung piping systems as long as the size of the piping is increased to account for the increased pressure drop due to MIC and the pipe wall thickness is increased to account for the anticipated rate of pitting corrosion. (See Chapter 7.)

6.2   FOULING IN COOLING WATER PIPING

6.2.1   BACKGROUND

Biofouling is the undesirable accumulation of biotic deposits on a surface. The deposit may contain micro- and/or macro-organisms. Microfouling is the development of a biofilm such as pseudomonas, on the surface in conjunction with inorganic particles such as suspended solids, scale, or corrosion products.[21] As such, microfouling is often a precursor to MIC. Macrofouling is the existence of biological creatures such as mollusks or aquatic weeds such as seagrass or milfoil. Suspended solids such as mud and sand or fish such as gizzard shad are also sources of fouling in cooling water systems.

Biofouling that is often prevalent in cooling water systems can be the source of many problems. Microfouling per se is not normally an impediment to flow as is corrosion, but when it appears on the inside diameter of HXs, heat transfer is impeded. Macrofouling is often the more serious impediment to flow, as it can not only block the entrance to HX tubes but also obstruct the IPS. Suspended solids can also be a problem in HXs, normally collecting and sometimes completely blocking the lower tubes in a HX.

For many years, fouling had been acknowledged to be an inevitable problem in cooling water systems, but with the advent of commercial nuclear power plants, fouling became unacceptable when it challenged the ability of nuclear power plants to shut down safely. (See Section 1.8.) The result was an industry-wide effort to identify biofouling and develop solutions to the problem. Some biofouling problems such as fish, seagrass, and sediment are quite site specific, while others like microfouling and macrofouling such as certain mollusks are quite widespread and are treated in more detail herein.

6.2.2 ASIATIC CLAMS

The Asiatic clam (corbicula fluminae), an exotic freshwater bivalve mollusk, was first discovered in North America near the mouth of the Columbia River in 1938[22] and has since spread throughout the United States. Unless precautions are taken, Asiatic clams often enter the cooling water system as adults or larvae when the system was initially flooded or soon after it began operation. Once in a cooling water system, Asiatic clams can grow rapidly, as much as 1/10-inch (0.254 cm) per month, and may grow to be 2 inches (5.1 cm) in size and can produce veligers throughout the year.[23] Adult Asiatic clams cannot attach to a surface and have limited mobility with a single "foot", so they are frequently found in low areas or areas of low velocity. As documented in Section 1.8, they have been a major concern to the NRC, as they have been the cause of numerous challenges to safety-related cooling water systems that are used only infrequently. The clams settle in upstream areas and grow until the system is called upon to operate. They are then sometimes swept up in the cooling water flow until they obstruct an important piece of equipment such as a valve or HX. Asiatic clams elude periodic chemical treatment regimens, as they sense the presence of the chemical and can close up until the danger passes. However, veligers smaller than 1/32-inch (0.079 cm) have not yet developed a protective shell and are thus susceptible to biocides. Asiatic clams spawn when the cooling water temperature is above 60°F (15.6°C) and have been reported to spawn well below 65°F (18.3°C).[23]

6.2.3 ZEBRA MUSSELS

Unlike the Asiatic clam, the zebra mussel's byssal threads can attach to almost any surface.[24] Zebra mussels have spread rapidly, as their planktonic (drifting) veliger stage remains in the water for several weeks after release, allowing it to be carried great distances before settlement. By 1991, zebra mussels were found in the TVA's Kentucky Lake near where the Tennessee, Ohio, and Mississippi Rivers converge. In 1994, EPRI predicted that zebra mussel infestations would eventually be found in 70% of the states. Presently, they may be found in 28 states or over 50% of the states. Zebra mussels can be found where the pH is greater than 5.5 and the water temperature is between 54 and 90°F (12.2 and 32.2°C).[24] Since they feed on suspended solids, zebra mussels have brought about great change in the ecology of bodies of water such as the Great Lakes, resulting in a shift in species composition due to the increasing clearing of the water.[25]

Unlike the Asiatic clam, the zebra mussels attach to hard surfaces by byssal threads to almost any surface.[26] The zebra mussel can grow up to 2 inches (5.1 cm) in length[27] and are a serious hazard to cooling water systems, as they may be found on IPS trash racks and forebays, system piping, and HXs.[24] Fortunately, zebra mussels may be controlled by a combination of monitoring and chemical treatment or heat treatment. Sodium hypochlorite in doses of less than 1.0 ppm is effective in controlling zebra mussels.[28] However, areas that may be up stream of the biocide injection point such as IPS trash racks and forebays pose a challenging problem.

6.3 CHEMICAL AND HEAT TREATMENT

6.3.1 CHEMICAL TREATMENT

Continuous chlorination with sodium hypochlorite of less than 1.0 ppm is widely used at power stations to control mollusks. Chlorination dosing at low levels not only stops the settlement of larvae but also reduces drastically the growth of microfouling and kills or removes the biofouling in the long term more effectively than is achieved with intermittent dosage.[22] Chlorine concentration tests indicate that there is hardly any difference between chlorine concentrations of 0.5 and 1.0 ppm in controlling biofouling.[25] Continuous or long-term molluscicide treatment is more effective than short-duration intermittent treatment because of the mollusks' inability to avoid long-term exposure.[29] However, chemical controls are more effective at preventing the settlement and growth of mollusks than mitigating the effects of established colonies. No chemicals have proved able to dissolve or remove the byssal threads which zebra mussels use to attach themselves to the IPS and piping.[24]

Since 1979, TVA has had good experience controlling Corbicula, except where mechanical or operational problems were experienced that interrupted chlorination.[30] Bleach feed systems are often plagued with frequent down time due to the corrosivity of the chlorine bleach and the pumps becoming air locked from off-gassing of the bleach. Problems with the equipment resulted in long periods of no treatment and some periods of over treatment.[31] TVA recommended that all incoming water to the cooling water systems should be strained. Straining is performed by automatic backwash type straining units located immediately upstream or downstream of the main pumping units of the system (i.e., at the source). Strainers have 1/32-inch (0.079 m) wire mesh baskets and are designed for periodic or continuous backwashing. TVA initially elected to use sodium hypochlorite generated onsite as needed to be injected upstream or immediately downstream of the strainers to kill the veligers before they get a chance to develop protective shells. Secondary water sources (such as jockey pumps, normally open interconnections with other water systems, etc.) also should be strained and chlorinated. If the incoming water has already been chlorinated, no additional injection is necessary. The chlorine level throughout most cooling water systems and at the system discharge is to be maintained at a total residual chloride (TRC) of 0.6 to 0.8 ppm during the entire clam spawning period. The clam spawning period as defined here is that period of time when the system inlet temperature normally exceeds 60°F (15.6°C).[30]

System design must be considered in conjunction with plant chemical discharge limits in defining the exact location of chemical injection. It should be remembered that the cooling water systems that are chlorinated often constitute only a small flow/ volume in comparison with the CCW system in a power plant and if the cooling water discharge of the system is mixed with CCW before being discharged to the environment, the impact of the chlorine on the environment may be minimal.

For safety-related cooling water systems, chlorination should be continuous to protect butt-welded SS piping from MIC attack by depriving it of pseudomonas bacteria that enable MIC to initiate at the welds. Additionally, veligers that enter the plant in the winter can grow shells before chlorination resumes. Experience has shown that when the chlorination is interrupted even though the cooling water is strained, control of mollusks may be lost. Once the veligers settle in the system, they can grow a shell and when chlorination is resumed, they are able to close and avoid being killed.[30]

6.3.2   COATINGS FOR ZEBRA MUSSEL CONTROL

Laboratory tests of silicone elastomer-based "foul release" coatings have shown favorable performance over a 24-month test period for the Diablo Canyon power plant located on the Pacific Ocean.[31] Two coatings containing toxic compounds that were deemed acceptable for use in the marine waters of Long Island Sound, New York, were cuprous oxides and copper epoxies. The single nontoxic coating found to be effective is silicone rubber impregnated with silicone oil expected to last five years. The silicone coating was also expected to be effective in controlling zebra mussels in freshwater environments.[32]

6.3.3   HEAT TREATMENTS FOR ZEBRA MUSSEL CONTROL

European research over the years has identified several lethal combinations of temperatures and durations for zebra mussels ranging from 91°F (32.8°C) for five hours to 97°F (36.1°C) for one hour with instant mortality at 104°F (40.0°C).[27] This range of intake temperatures is confirmed by experience in the United States where a temperature of 92°F (33.3°C) for 2.5 hours was found to be 100% effective.[33] At Detroit Edison's Harbor Beach power plant, an oxygen scavenger (sodium sulfide) was used in conjunction with thermal treatment to asphyxiate and heat zebra mussels over a several day period.[34] The heating of cooling water has been accomplished by recirculating CCW and by the use of auxiliary boilers.

6.4   TVA QUALIFICATION TESTING OF CHEMICAL TREATMENT

6.4.1   CORROSION INHIBITOR TESTING

Tests were conducted at SNP in order to make the results directly applicable to the SNP ERCW system. Zinc polyphosphate was tested by TVA as the type of corrosion inhibitor most likely to be both environmentally acceptable and cost-effective in their applications. Recognizing that cooling water systems were to be chlorinated for Asiatic clam control, sodium hypochlorite was tested as a potential corrosion inhibitor along with zinc polyphosphate. The test stand included a series of three 1-foot long (0.305 m), 1-inch (2.54 cm) CS spool pieces arranged in parallel, each carrying approximately 15 gal/min (0.95 l/s). Zinc polyphosphate was injected into one line at approximately 1.5 ppm. Sodium hypochlorite was injected into the second line at a rate intended to yield a chlorine residual of 0.6 to 0.8 ppm at the outlet of the line. The third line served as a control with no treatment of the cooling water. A sample spool piece of pipe was removed from each of the three lines on four occasions over a period of about one year. The volume reduction of the interior of each sample was measured as described in Section 6.2.1 to determine the extent of corrosion products buildup.[1]

The sample spool pieces were removed after 1, 3, 6–1/2, and 11–1/2 months of service, and results are plotted in Figure 6.14.[1]

The samples taken at one month and at three months showed rapid accumulation of deposits. Later samples showed a slower accumulation and in some cases, an apparent reduction in deposits. All of the test samples had numerous small tubercles which were similar in appearance to one another and to the large tubercles observed in the sampling program described in Section 6.4. The earlier samples had numerous tubercles up to about 1/16-inch (0.159 cm) high but also had sizeable areas which were completely covered with tubercles up to about 1/8-inch (0.318 cm) high. The average diameter reduction in the later samples was on the order of 1/16-inch (0.159) in the untreated line and the corrosion inhibited line and on the order of 1/10-inch (0.254 cm) in the chlorinated line.[1]

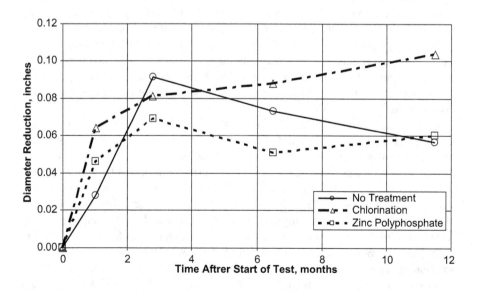

FIGURE 6.14 Results of corrosion inhibitor test as measured by diameter reduction.

The observed high initial rate of buildup and lower rate later in the test are consistent with the sample data described in Section 6.1.2 and plotted in Figure 6.4. The study concluded the following:

- The rate of buildup was initially very rapid but had slowed dramatically by the end of the test.
- The chlorinated line experienced more rapid buildup of deposits than the other two lines.
- Zinc polyphosphate in the concentrations tested had little if any beneficial effect under the test conditions.
- Corrosion-resistant materials would be needed in many instances instead of CS because of the serious corrosion problems with CS and the lack of effectiveness of corrosion inhibitors.[1]

6.4.2 Biocide Testing

Prior to implementing the proposed cooling water system chemical treatment regimen, TVA sponsored testing by various organizations to ensure the feasibility of the proposed regimen. The questions that needed to be answered were as follows:

- Is the regimen harmful to the environment?
- Will the regimen damage the cooling water system components?
- Is the regimen efficacious?

Toxicity tests were conducted by the Ecology Laboratory of the Oak Ridge National Laboratory (ORNL) to provide the necessary technical information to support regulatory approval for the use of the biocide in the ERCW system at SNP. The biocide tested was Actibrom 1338, a proprietary bromine and dispersant mixture manufactured by Nalco Corporation. The tests were conducted with the Actibrom 1338 mixed with sodium hypochlorite at a 1:1 molar ratio. Tests conducted included both a Ceriodaphnia seven-day survival and reproduction test and a fathead minnow larval growth test also of seven days' duration. The Actibrom 1338/sodium hypochlorite mixtures (1:1 molar ratios) did not result in increased mortalities at concentrations in excess of 30 ppm.

Qualification tests were conducted by the SM&E Laboratory to evaluate the potential adverse effects of candidate biocides on the material of construction of the wetted parts of a typical cooling water system and to ensure that the introduction of the biocide into the system does not cause the corrosion products buildup on the CS piping to become detached from the surface of the pipe wall. A test was conducted in which three samples of 2.5-inch (6.4 cm) CS pipe and three samples of 6-inch (15.2 cm) SS pipe all containing corrosion products including tubercles were subjected to a 30-day exposure. One sample each of carbon and SS was exposed to untreated raw water, one each was exposed to a raw water solution treated with a solution of generic sodium bromide activated by sodium hypochlorite, and one each was exposed to raw water treated with Nalco Actibrom 1338 mixed with sodium hypochlorite. The biocides were initially injected at 7.5 ppm of free available chlorine with a 1:4 molar ratio of bromine to chlorine. A visual and tactile inspection of the SS specimens at the end of the test revealed an absence of slime on the surfaces of biocide-treated specimens, but no significant amount of corrosion products was removed from the CS specimens.[35]

Corrosion tests were conducted on materials typically found in ERCW systems to determine the relative corrosivity of candidate biocides on these materials in a sterile solution of simulated raw water. The primary purpose of this test was to determine if the addition of bromine or bromine with a dispersant increases the corrosion rate when compared with sodium hypochlorite alone. The test was accelerated somewhat by employing a biocide level of 7.5 ppm, which was five times the anticipated level. Biocides containing bromine did not appear to be more corrosive than sodium hypochlorite alone. In fact, the addition of commodity sodium bromide resulted in a corrosion rate decrease for all CS specimens tested, accompanied by an increase in the amount of passivated surface area. The slight increase in the corrosion rate with Actibrom 1338 appeared to be due to the dispersant in this solution acting to remove corrosion products from the surface of the CS. All copper alloys tested exhibited good protection in the simulated raw water without biocide and significant corrosion rates when exposed to sodium hypochlorite. All of the copper alloys showed a decrease in corrosion rate when bromine was present in the biocide. All of the austenitic SS specimens and weld coupons exhibited very low corrosion rates. However, all of the austenitic SSs except AL-6XN exhibited some susceptibility to crevice attack in one or more of the solutions, and some of the austenitic stainless welded coupons exhibited minor crevice attack in the solutions containing bromide.[36]

A test was conducted to evaluate the potential degradation effects on Buna-N rubber materials which are used as O-rings in the ERCW system. Some of the O-rings tested were sandwiched between corrosion coupons used in the test described above while others were stretched over the coupon assemblies to provide the compressive force for the sandwiched O-rings. Physical property measurements were selected for their response to potential degradation, due to exposure to the biocides and included hardness, tensile strength, elongation, and modulus. The results of these tests, performed in accordance with the appropriate industry standard, showed no physical property degradation as a result of exposure to the biocide.[37]

Tests were conducted by the University of Tennessee's Institute for Applied Microbiology (IAM) to determine the effectiveness of various biocides at controlling sessile and/or planktonic bacteria populations and on controlling MIC. An initial test was conducted on MIC nodule on a 6-inch (15.2 cm) SS weld removed from the SNP to determine if a biocide consisting of chlorine and a dispersant could penetrate the nodules and provide growth inhibition of the corrosive microorganisms without dissolution or removal of the nodule. The test was conducted in a re-circulating system in which the test water was replaced with fresh lake water treated with biocide at approximately 48-hour intervals to maintain a total available chlorine concentration of approximately1 ppm. During the test, the pH varied between 7.4 and 8.75. No significant reduction in the size of the nodules was observed. Biological tests conducted by IAM concluded that although there remained a very diverse population of organisms in the nodules, their numbers were very low and those that were present were in a dormant condition.[38]

A series of tests was conducted by IAM which were specifically designed to determine the biocide composition, dosage, and application time required to be effective controlling MIC-causing bacteria.[38] In six identical test stands, CS and SS specimens were exposed to a re-circulating sterile medium. Three of these test loops were inoculated with a mixed consortium of six types of bacteria including FeRed, FeCcpt, SRB, and Ps taken from TVA nuclear plants. Within a few days, sessile bacterial colonies were formed as tubercles on the CS specimens in the inoculated loops (tubercles did not form on the SS specimens). After the colonies were established, two of the three inoculated loops were treated with a biocide, with the third serving as a control. The first test loop was treated with NaOCl alone, the second was treated with commodity NaBr/NaOCl. The same biocides were added to two of the sterile loops with the third sterile loop left as a control. Direct electrical measurements were made of the corrosion occurring in mohms, and microbes were recovered after each test and the number of colony-forming units (CFU) was determined. The activity of the biofilms was determined by measuring the C-14 acetate uptake by the microbial lipids on the specimens.

Two separate experiments were conducted. In the first experiment, both biocides were fed to maintain a residual halogen of 2 ppm for one hour. This was followed in 24 hours with an additional exposure of 2 ppm for two hours. The NaBr/NaOCl biocide was mixed at a 1:4 molar ratio and pH was maintained at 8.5. This treatment regimen proved to be totally ineffective for both biocides. At the end of the second biocide injection, the corrosion rates for the inoculated specimens that were treated with biocides were higher than the control, and the viable organisms recovered from the specimens showed no evidence of damage to the bacteria and in many cases there was an increase in viable cells. These results suggest that ineffective or sub-lethal pulses of biocide actually stimulate the bacteria. Each of the specimens, with and without biocide additions, developed a corrosion products layer over the specimen which when removed revealed a smooth surface. In contrast, those specimens that were inoculated with bacteria exhibited severe pitting on the metal surface after removal of the biofilm.

A second experiment was designed in which both biocides were fed at slug dose of 16 ppm for two hours followed by a continuous residual halogen of 2 ppm for 24 hours. The NaBr/NaOCl biocide mix was changed to a 1:1 molar ratio, and the pH was kept at 8.5. This second treatment regimen had a marked biological effect. Although the presence of oxidants in the system increased the general corrosion rate both with and without bacteria present, the test results shown on Figure 6.15 indicate that both biocides were effective in killing bacteria, and that the NaBr/NaOCl biocides may be slightly more effective on sessile (coupon) bacteria than NaOCl alone. This result is to be expected since the pH for the test was selected to be prejudicial against chlorine alone. The data also suggest that both biocides are effective against a broad spectrum of bacteria. Figure 6.15 suggests that although a biocide may lower the number of CFUs below 10^6/ml, thus rendering the biofilm inactive, even bacteria that were suppressed below detectable levels were able to partially recover within 24 hours after termination of biocide treatment. Micrographs

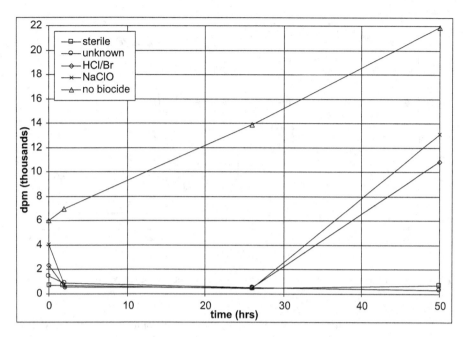

FIGURE 6.15 Activity of biofilm measured as C14-acetate.

of the specimens tested showed that pitting corrosion characteristic of MIC is greatly reduced when an effective biocide is employed support the numerical data.[38]

6.5 TVA EXPERIENCE WITH CHEMICAL TREATMENT

6.5.1 SEQUOYAH NUCLEAR PLANT

The ERCW system at TVA's SNP employs automatic backwash strainers as depicted in Figure 3.6 near the discharge of each of the 8 ERCW pumps. These strainers filter out objects larger than 1/32-inch (0.079 cm). SNP was the first TVA nuclear plant to implement a biocide treatment of their safety-related ERCW system. As discussed in Section 6.3.1, continuous biocide treatment is important. For example, in March 1982, while conducting a surveillance test of the ERCW system at SNP, a flow decrease to the containment spray HX from 100% to 31% of rated capacity was noted. The pipe was opened upstream of the manual inline strainer, revealing approximately 15 gallons (0.0568 m³) of clam shells that were restricting the flow. Under normal operating conditions, the 18-inch (45.7 cm) header supplying the HX is stagnant except for a 1-inch (2.54 cm) mini-flow line around the HX. The mini-flow line was found to be clogged. The ERCW was strained but only periodically chlorinated in the summer of 1981. This condition was conducive to clam growth. Subsequently, steps were taken to ensure flow through the mini-flow line and chlorination by repairing the hypochlorite injection systems. Continuous chlorination as described above was practiced in 1982 with complete control of Asiatic clams.[30]

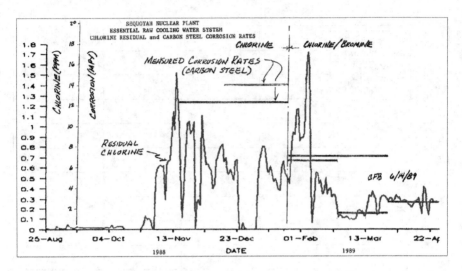

FIGURE 6.16   Halogen residual and corrosion rates in the SNP ERCW system.

Injection of the chlorine/bromine biocide Nalco Actibrome 1338 began at SNP in January 1989. Figure 6.16 shows the residual halogen and corrosion rates before and after the change. As one may see, shortly after the change to Actibrome 1338, the corrosion rate in the system as measured by corrosion coupons and the required residual halogen dropped dramatically. In addition to greatly reducing the CS corrosion rate, the biocide was very effective in cleaning up the system so that there is less demand for oxidant while still maintaining a free, available halogen residual.

Although qualification tests conducted by the SM&E Laboratory concluded that the introduction of a biocide into the system would not cause the corrosion products buildup on the CS piping to become detached from the surface of the pipe wall,[35] one might conclude from the results of its implementation at SNP that the biocide gradually cleans up the system. The improvement is likely due to not only the hyprobromous acid but also the dispersant in the Actibrome 1338. As discussed in Section 7.4, flow and pressure drop tests conducted on the ERCW system headers between the IPS and the auxiliary building support the suggestion that biocide treatment over a long period of time tends to clean up the system, since the pressure drop in the piping decreased.[39,40] However, plant personnel reported a deposit of iron oxide and silt in the test water box and on the tubes of some HXs that was not there on prior inspections, which may result in the loss of heat transfer capability.[41]

6.5.2   WATTS BAR NUCLEAR PLANT

The ERCW system at TVA's WBNP employs automatic backwash strainers as depicted in Figure 3.6 near the discharge of each of the 8 ERCW pumps. These strainers filter out objects larger than 1/32-inch (0.079 cm). WBNP injected sodium hypochlorite on a continuous basis any time the temperature of the cooling water was above 60°F. In December of 1992, TVA implemented a chemical treatment program

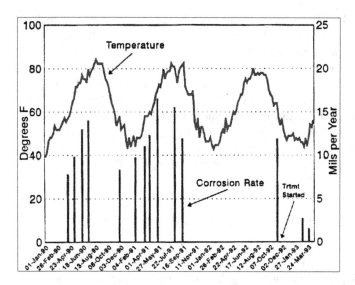

FIGURE 6.17 WBNP carbon steel corrosion rate vs. cooling water temperature.

at WBNP consisting of bromochlordimethylhydantoin (BCDMH), a granular form of bromine/chlorine, used to reduce the concentration of bacteria that can cause MIC. BCDMH was chosen because it is more effective at pH > 7.5 and is a dry chemical that is safer to work with and is a more reliable feed system. Cleanup of the ERCW system was accomplished by continuous application of polyphosphate to soften and remove MIC nodules over time and a copolymer dispersant to maintain sequestered small particles in suspension to be removed by the flowing water. As shown in Figure 6.17, CS corrosion rates were controlled by continuously injecting a zinc-based corrosion inhibitor in addition to the polyphosphate. The chemicals were injected into the IPS in front of the traveling screens to eliminate problems with overcoming the high ERCW pump discharge pressure.[41]

The BCDMH was injected continuously for three weeks twice each year during the clam spawning season. DGH-QUAT, a non-oxidizing biocide, was used for 12 to 24 hours on a quarterly basis to flush out normally stagnant lines to control Asiatic clams and zebra mussels.[41]

6.5.3 BROWNS FERRY NUCLEAR PLANT

At BFNP, the strainers at the discharge of the cooling water pumps strain to 1/8-inch (0.318 cm) which permits veligers that have already developed a shell to enter the system. Sodium hypochlorite was fed at BFNP with a target TRC of 0.2 to 0.8 ppm when the cooling water temperature exceeded 50°F (10°C) to control Asiatic clams and MIC in butt welds. The hypochlorite feed system was frequently out of service. Problems with the equipment resulted in long periods of no treatment and some periods of overtreatment with residuals as high as 6 ppm. Despite the treatment program, Asiatic clams continued to foul HXs. In 1995, BFNP implemented a different chemical treatment program for the cooling water systems. For biological control, bromine was added to sodium hypochlorite to produce hyprobromous acid and fed four hours/day with a halogen residual (0.1–0.5 ppm). One reason that bromine activated by sodium hypochlorite was preferred over straight chlorine is that it is less corrosive than chlorine. Other advantages include better performance at elevated pH and less persistence in the environment. However, hypobromous acid fed for a few hours at a time has very limited effect on macrofouling by clams and mussels. The chemical treatment program also included phosphate (1–1.8 ppm), zinc (0.2–0.25 ppm), and Copolymer (0.28–0.35 ppm) fed continuously and Azol blend (3.5–4.0) 15 minutes 1–2 times/month, DMAD (0.5–0.62) fed for 30 minutes once a week, and Quat (0.15–1.5 ppm) fed for 24 hours twice a week. BFNP reported that the results were as good or better than expected.[42]

6.6   CONCLUSIONS AND RECOMMENDATIONS

The cooling water system is perhaps among the most poorly engineered systems in a power or industrial plant. Most employ CS piping and make little or no provision for controlling corrosion or fouling. Unless a strict chemical regimen is observed at the outset, CS piping is not an appropriate material for small bore piping and often is not suitable for large bore piping. As shown in Section 6.1, pitting corrosion rates for CS piping due to MIC can be as high as 16 MPY or more, leading to through-wall leaks in as few as 23 years in standard 3/8-inch (0.9525 cm) wall piping for pipe sizes over 12 inches (30.5 cm) and even sooner for smaller line sizes. Further, as shown in Chapter 7, the flow-passing capability of CS piping is greatly reduced due to the tuberculation associated with MIC that increases the roughness factor and reduces the effective inside diameter of the pipe. Studies have shown that in terms of the flow-passing capability of the piping, the installed cost of SS piping (the preferred material for hung piping) may be approximately the same as for CS. However, butt-welded SS piping may be vulnerable to MIC attack at the weld joint, which can lead to leaks and loss of structural strength over time. Therefore, a biocide treatment is indicated where SS piping 4-inch (10.2 cm) and larger is employed.

A very large percentage of power and industrial plants are subject to fouling, both microfouling that initiates MIC and reduces heat transfer in HXs and macrofouling that obstructs flow. Straining of cooling water at the source is essential to a successful biocide treatment program. Particles larger than 1/32-inch (0.079 cm) should be strained out with automatic backwashing strainers such as the S. P. Kinney Model A strainer with wire baskets is shown in Figure 3.6. Veligers smaller than 1/32-inch (0.079 cm) have ciliated flaps for swimming and feeding but have not yet developed hard shells allowing them to close in the presence of a biocide. Therefore, they may be killed instantly with a low dose of an oxidizing biocide such as chlorine. Failure to adequately strain cooling water makes controlling mollusks almost impossible, as they can sense the presence of a biocide and close up their shells.

A biocide system is essential to managing most fouling problems such as mollusks. The cheapest biocide is chlorine, normally sodium hypochlorite (bleach) a liquid that is highly corrosive and often difficult to handle. Bromine is sometimes mixed with chlorine to make hyprobromous acid. The advantage of bromine is that it is less corrosive than chlorine (especially with copper-alloyed tubes), is more effective at pH > 7.5, and is less persistent in the environment. However, if the residuals are kept low and the cooling water is mixed with the CCW before being discharged, a chlorine-only biocide is not normally offensive. Commercial biocides such as Nalco's Actibrome 1338 also contain a proprietary dispersant. Actibrome 1338 has been shown to clean up corroded CS piping over time. However, the result may be a relocation of the iron and other constituents on the surfaces of HX tubes.

Handling and mixing liquid chlorine and bromine in the right proportions can be challenging. In some locations where pH < 7.5, one may be able to achieve satisfactory results with just sodium hypochlorite alone or in conjunction with a polyphosphate to soften and remove MIC nodules over time and/or a copolymer dispersant to maintain sequestered small particles in suspension to be removed by the flowing water if required.

By far the greatest impediment to a successful biocide treatment program is the reliability of the chemical feed system. Adequate engineering must go into the design of these systems to ensure reliable operation. Consideration should be given to installing redundant components such as pumps, etc., or even full redundant trains of equipment to achieve the required 100% availability. This job is often left to the chemical manufacturer without adequate oversight.

Common practice is to suspend biocide treatment except during the mollusk spawning season or to rely on periodic shock treatments. Biocide treatment programs that rely on intermittent operations permit veligers to enter the cooling water system where they hide out and grow the protective shell. Even if the periodic treatment is of such duration as to starve the mussel, the result is a dead mussel in the system, which is hardly better than a live one.

In areas where zebra mussels are a problem, one should control their growth upstream of the point of biocide introduction. The use of antifouling paints such as cuprous oxides, copper epoxies, or nontoxic coatings such as silicone rubber impregnated with silicone oil should be considered. The system at WBNP of introducing BCDMH (commonly used in swimming pools) upstream of the trash racks and/or traveling screens is of interest in that it holds promise for controlling zebra mussels as well as Asiatic clams. One might surmise that this arrangement could lead to accelerated corrosion of the traveling water screens if they are made of CS.

Unlike oxidizing biocides such as chlorine or bromine where the residual becomes relatively inert when their demand is satisfied, non-oxidizing biocides such as Clamtrol or DGH-QUAT are poisons that are long-lasting and very expensive. Their use in open cooling-water systems is highly problematic, generally requiring the system to be shut down, isolated, and flushed after the mollusks have been killed.

The perfect is often the enemy of the good. In some cases, the cost of the chemical treatment system has become so high as to become unaffordable, resulting in the elimination of chemical treatment all together. Care should be taken to implement the least costly system that is effective. In most instances, this will be determined by site-specific conditions.

REFERENCES

1. Bain, W. S., et al. *Corrosion in Carbon Steel Raw Water Piping*, (TVA Ref. ESS800416204) Tennessee Valley Authority, Knoxvville, TN, 1979.

2. Bowman, C. F. and W. S. Bain. A New Look at Design of Raw Water Piping. *Power Engineering Magazine*, Vol. 84, No. 8, pp. 73–77, August 1980.

3. TVA Mechanical Design Guide DG-M5.2.1. *Corrosion/Erosion Allowance for Determination of Minimum Pipe Wall Thickness in Carbon Steel Piping Systems*, Tennessee Valley Authority, Knoxville, TN, 1980.

4. IN 85-30. *Microbiologically Induced Corrosion of Containment Service Water System*, Nuclear Regulatory Commission, Washington, DC, 1985.

5. Generic Letter 89-13. *Service Water System Problems Affecting Safety-Related Equipment*, Nuclear Regulatory Commission, Washington, DC, 1989.

6. "Microbiologically Influenced Corrosion (MIC)", Institute of Nuclear Power Operations Significant Event Report SER 73–84, 1984.

7. Licina, George. *Sourcebook for Microbiologically Influenced Corrosion in Nuclear Power Plants*, Electric Power Research Institute, Palo Alto, CA, 1988.

8. Pavinich, W. A. et al. *Four-Point Bend Testing of MIC-Damaged Piping*, (TVA Ref. B45890825251) Tennessee Valley Authority, Knoxville, TN, 1989.

9. Deardorff, A. F., et al. Evaluation of Structural Stability and Leakage from Pits Produced by MIC in Stainless Steel Service Water Lines. *Corrosion 89*, 1989.

10. Pavinich, W. A. et al. Mechanical Properties of Stainless Steel Weldments Degraded by MIC, Paper No. 531. *Corrosion 90*, 1990.

11. *TVA Civil Design Standard DS-C1.2.8, Structural Evaluation of Microbiologically Induced Corrosion Degradation in Piping*, Tennessee Valley Authority, Knoxville, TN, 1991.

12. Blackburn, F. E., and Mullin, L. J. *Detection and Control of Bacterial Corrosion Using Internal Corrosion Monitors. Microbially Influenced Corrosion and Biodeterioration, MIC Consortium*, Knoxville, TN, 1990, pp. S1–S24.

13. Bowman, C. F. and P.V. Guthrie, Jr. *Corrosion in Carbon Steel Service Water Piping. 1994 ASME Pressure Vessel and Piping Conference*, ASME, 1994.

14. Tuovinen, J. E., et al. Bacterial, Chemical, and Mineralogical Characteristics of Tubercles in Distribution Pipelines. *Journal AWWA*, American Water Works Association, Denver, CO, pp. 626–635, November 1980.

15. Metell, H. M. *Corrosion from Iron Bacteria in Plant River Water Systems. Proceedings, Workshop on Microbe Induced Corrosion*, Electric Power Research Institute NDE Center, Charlotte, NC, 1986.

16. ANSI/ANS-59.1-1986. Nuclear Safety Related Cooling Water Systems for Light Water Reactors. *American Nuclear Society*, 1986.

17. Generic Letter 90-05. *Guidance for Performing Temporary Non-Code Repair of ASME Code Class 1, 2, and 3 Piping*, Nuclear Regulatory Commission, Washington, DC, 1990.

18. ASME B31G-1991. *Manual for Determining the Remaining Strength of Corroded Pipelines*, American Society of Mechanical Engineers, New York, NY, 1991.

19. Herro, H. M. *Corrosion and Fouling in Nuclear Power Plant Service Water Systems, Electric Utility Service Water System Reliability Improvement-A Compendiun of Presentations*, Electric Power Research Institute, Palo Alto, VA, 1993.

20. Herro, H. M. and R. D. Port, *The NALCO Guide to Cooling Water Systems Failure Analysis*, McGraw Hill, Inc., New York, 1993.

21. Characklis, W. G., et al. *Biofouling Control Technology: The Roll of Fouling Monitors. Condenser Technology Symposium*, Electric Power Research Institute, Providence, RI, 1987.

22. Jenner, H. A., et al. *Hydroecologie Appliquee*, Tome 10, Volume 1–2, Electricite de France, 1998.

23. Derrworth, J. E., et al. *Duke Power Company's Development of a Biofouling Monitoring Program at Two Nuclear Power Plants on the Southeastern Reservoirs. International Macrofouling Symposium*, Electric Power Research Institute, Orlando, FL, 1990.

24. Mussalli, Y. G. *Zebra Mussel Monitoring and Control Guide*, Electric Power Research Institute, Palo Alto, CA, 1992.

25. Jenner, H. A. *Biomonitoring in Chlorination Anti-Fouling Procedures to Achieve Discharge Concentrations as Low as Reasonable. International Macrofouling Symposium*, Electric Power Research Institute, Orlando, FL, 1990.

26. Armor, A. *Zebra Mussels: The EPRI Response Update. International Macrofouling Symposium*, Electric Power Research Institute, Orlando, FL, 1990.

27. DeMoss, D., et al. *Design Features to Mitigate Fish, Zebra Mussel, and Other Macrofouling of Plant Components. International Macrofouling Symposium*, Electric Power Research Institute, Orlando, FL, 1990.

28. Matisoff, G., et al. *Controlling Zebra Mussels at Water Treatment Plant Intakes II - Velinger Dose/Response Static Tests. International Macrofouling Symposium*, Electric Power Research Institute, Orlando, FL, 1990.

29. Garbarino, L. N. *Testing of Chemical Methods for Clam Control under Simulated Field Conditions. International Macrofouling Symposium*, Electric Power Research Institute, Orlando, FL, 1990.

30. Isom, B. G. et al. Controlling Corbicula (Asiatic Clams) in Complex Power Plant and Industrial Water Systems. *American Malacological Bulletin*, Special Edition No. 2. pp. 95–98, 1986.

31. Sommerville, D. C. and F. L. Steinert. *Development of Alternative Macrofouling Control Methods for the Diablo Canyon Power Plant. International Macrofouling Symposium*, Electric Power Research Institute, Orlando, FL, 1990.

32. Gross, A. C. *Macrofouling Problems and Control Techniques in Marine water around Long Island. International Macrofouling Symposium*, Electric Power Research Institute, Orlando, FL, 1990.

33. Kahabka, J. *The Zebra Mussel: New York's Experience. International Macrofouling Symposium*, Electric Power Research Institute, Orlando, FL, 1990.

34. Harwood, D. B. and D. J. Buda. *Effective Zebra Mussel Control at Detroit Edison Harbor Beach Power Plant. American Power Conference*, Chicago, IL, 1993.

35. Guthrie, P. V., et al. *Sequoyah Nuclear Plant Dipersion Tests on Carbon and Stainless Steel Pipe Samples*, Test 6 (R3), SME-COR-88-001, Singleton Materials Engineering Laboratory, 1988.

36. Guthrie, P. V., et al. *Sequoyah Nuclear Plant Evaluation of the Effects of Cl/Br Biocides on the Corrosion of Metals within the ERCW*, SME-COR-88-011, Singleton Engineering Laboratory, 1988.

37. Guthrie, P. V., et al. *Sequoyah Nuclear Plant Evaluation of the Effects of Cl/Br Biocide Water Treatment on Buna-N Rubber*, SME-COR-88-019, Singleton Engineering Laboratory, 1988.

38. Van, A. et al. *Accelerated Tests of Effects of Bromide Additions to Biocide Treatments on Microbiologically Induced Corrosion (MIC) on Mild Steel*, Institute for Applied Microbiology, 1988.

39. Bowman, C. F. Solving Raw Water Piping Corrosion Problems. *Power Engineering Magazine*, Vol. 98, No. 7, pp. 35–38, July 1994.

40. Hewette D. *Utility Experience Report – Tennessee Valley Authority. Service Water Working Group Meeting Proceedings*, Electric Power Research Institute, Palo Alto, CA, 1989.

41. Riggle, K. *TVA Watts Bar Nuclear Plant – Service Water Chemical Treatment Program, Service Water Systems Reliability Improvement Seminar*, Electric Power Research Institute NDE Center, Charlotte, NC, 1993.

42. Garbarino, N., et al. *Raw Water Treatment at the TVA Browns Ferry Nuclear Plant. Service Water Reliability Improvement Seminar*, Electric Power Research Institute, Orlando, FL, 1997.

7 Pipe Flow

7.1 EQUATIONS FOR HEAD LOSS BY PIPE FRICTION

7.1.1 DARCY-WEISBACH EQUATION

Equation (7.1) is the Darcy-Weisbach[1] equation for the friction head loss with turbulent flow in piping that is running full.

$$h_{L,f} = f \frac{L}{d_i} \frac{V_{pipe}^2}{2g} \tag{7.1}$$

where
$h_{L,f}$ = friction head loss
f = friction factor
L = length of pipe
d_i = inside diameter of pipe
V_{pipe} = water velocity in a pipe or tube
g = acceleration due to gravity.

Equation (7.2) is the value for the friction factor, f, determined from the Colebrook equation[2] as follows:

$$\frac{1}{\sqrt{f}} = - \, 2\log_{10}\left(\frac{e}{3.7 \, D} + \frac{2.51}{\text{Re} \, \sqrt{f}} \right) \tag{7.2}$$

where
e = absolute roughness of pipe, ft

Equation (7.1) is more commonly expressed in terms of volumetric flow rate in English units as follows:[3]

$$h_{L,f} = 0.0311 f \frac{LQ^2}{d^5} \tag{7.3}$$

where
Q = Flow rate, Gal/min
d = diameter of pipe, in.

7.1.2 HAZEN-WILLIAMS EQUATION

Whereas the Darcy-Weisbach equation provides a more analytically based mathematical solution to the head loss due to pipe friction, the empirical Hazen-Williams[4] equation is more appropriate for power plant cooling water system applications. As

95

shown in Chapter 6, the interior of cooling water piping systems is subject to deterioration with age depending on chemical properties of the water and pipe material in contact with the water. Scaling, corrosion, or bacterial attack can therefore adversely affect the flow capacity of the system. Some provision must be made in pressure drop calculations to account for this deterioration of the pipe interior surface. The most readily available information for a quantitative expression of pipe surface condition relative to operating experience and length of service life is the "C" factor used in the Hazen-Williams equation shown in Equation (7.4).[4]

$$h_{L,f} = 0.002083\,L \left(\frac{100}{C} \right) \left(\frac{Q^{1.85}}{d^{4.8655}} \right) \tag{7.4}$$

where
C = Hazen-Williams "C" factor

The Hazen-Williams formula is based on water at 60°F (15.6°C) having a kinematic viscosity of 1.13 centistokes. The kinematic viscosity of water can vary from 1.8 for water at 32°F (0°C) to 0.29 centistokes for water at 212°F (100°C), respectively. However, this variation in kinematic viscosity for the normal range of cooling water systems results in negligible changes in the friction head loss and may be ignored.

7.1.3 SELECTION OF ROUGHNESS COEFFICIENT

The main element of uncertainty in both the Darcy-Weisbach and Hazen-Williams methods of calculating the head loss due to pipe friction is the choice of the roughness coefficient (i.e. "e" Darcy-Weisbach and "C" for Hazen-Williams). Table 7.1 lists the recommended values to be used for the equivalent length from the indicated source.

TABLE 7.1
Commonly Used Roughness Coefficients for Design Purposes

Type of Pipe	C	e
Cement asbestos	130,[5] 140[4]	0.0075[5]
Bitumastic enamel-lined iron or steel centrifugally applied	140[4]	
Smooth glass or plastic	135,[5] 140[1]	0.005[5]
Cement-lined iron or steel centrifugally applied	130,[5] 140[4]	0.0075[5]
Cement mortar trawled in place	125[5]	0.011[5]
Copper, brass, lead, tin, or glass pipe and tubing	130[4]	
Welded and seamless carbon steel	55-70,[6] 100[1,4,5]	0.060[5]
Welded and seamless stainless steel	120[6]	
Wrought iron, cast iron	100[1,4,5]	0.060[5]
Tar-coated cast iron	100[4]	
Galvanized iron	90[5]	0.012[5]
Concrete, formed	80,[5] 100,[4] 130[1]	0.22[5]
Spiral-riveted steel (flow with lap)	100[4]	
Spiral-riveted steel (flow against lap)	90[4]	
Corrugated steel	60[4]	

7.2 PIPE FORM LOSSES

7.2.1 FORM LOSS EQUATION

Form losses are caused by changes in flow direction in valves and fittings, etc. in a piping system as opposed to those caused by friction at the pipe wall. Friction losses are calculated from the pipe centerline intersections of such components. Separate equations are necessary since form losses are not affected by changes in pipe wall roughness.[6] Form losses are independent of the properties of the fluid and of the pipe to which the valves and fittings, etc., are connected. Since the loss is a function solely of geometry and fluid velocity, coefficients (i.e., k factors) can be established that are valid for components of similar size and shape. Equation (7.5) is the equation relating the form loss to the "k" factor and the velocity of the cooling water passing through the component.[2]

$$h_{L,F} = k \frac{V^2}{2g} \tag{7.5}$$

where

$h_{L,F}$ = form head loss
k = resistance coefficient.

Tests have been conducted for most commercially available components of similar size and shape. The k factors for flanged pipe fittings shown in Figure 7.1 and valves in Table 7.2 are from Reference 5.

The "k" factor for other components such as strainers and HXs must be obtained from the manufacturer. (See Section 3.4 regarding S. P. Kenney strainers.)

7.2.2 EQUIVALENT LENGTHS

When performing head loss calculations by hand, common practice is to express the head loss through components in the cooling water piping system in terms of the equivalent length of pipe that would produce the same head loss. Table 7.3 shows the equivalent lengths of piping for an array of components from Reference 4.

The equivalent lengths were calculated for steel pipe carrying cold water. The error for water in a steel pipe at 100°F (37.8°C) is about -5%. The error for epoxy-coated or copper pipe is about 15% for water at 40°F (4.4°C) and 25% for water at 100°F (37.8°C). However, since these errors are in the conservative direction, they are commonly neglected.[2]

As seen in Reference 6 by F. P. Carr and C. F. Bowman, a separate form of the Hazen-Williams equation is required for form losses in piping components, since they are not affected by changes in wall roughness. Therefore, in order to use the equivalent length type analysis for form losses, the Hazen-Williams "C" factor associated with the published data for such losses must be inserted into the equation as seen in Equation (7.6).[6]

$$h_{L,F} = 0.001486 \left(\frac{Q^{1.85}}{d^{4.8655}} \right) \tag{7.6}$$

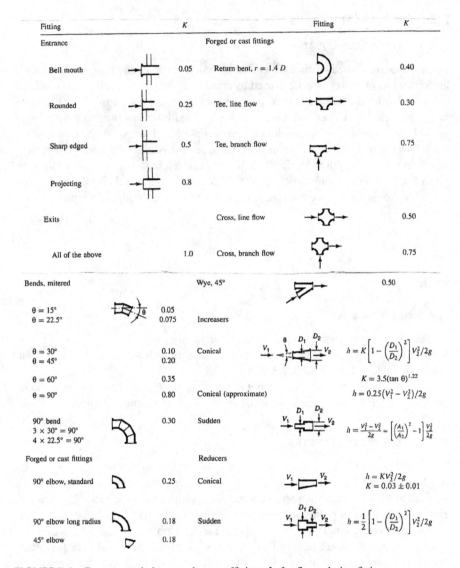

Fitting	K	Fitting	K
Entrance		Forged or cast fittings	
Bell mouth	0.05	Return bent, $r = 1.4\,D$	0.40
Rounded	0.25	Tee, line flow	0.30
Sharp edged	0.5	Tee, branch flow	0.75
Projecting	0.8		
Exits		Cross, line flow	0.50
All of the above	1.0	Cross, branch flow	0.75
Bends, mitered		Wye, 45°	0.50
$\theta = 15°$	0.05		
$\theta = 22.5°$	0.075	Increasers	
$\theta = 30°$	0.10	Conical	$h = K\left[1 - \left(\dfrac{D_1}{D_2}\right)^2\right] V_2^2/2g$
$\theta = 45°$	0.20		
$\theta = 60°$	0.35		$K = 3.5(\tan \theta)^{1.22}$
$\theta = 90°$	0.80	Conical (approximate)	$h = 0.25(V_1^2 - V_2^2)/2g$
90° bend	0.30	Sudden	$h = \dfrac{V_1^2 - V_2^2}{2g} = \left[\left(\dfrac{A_1}{A_2}\right)^2 - 1\right]\dfrac{V_2^2}{2g}$
$3 \times 30° = 90°$			
$4 \times 22.5° = 90°$			
Forged or cast fittings		Reducers	
90° elbow, standard	0.25	Conical	$h = KV_2^2/2g$ $K = 0.03 \pm 0.01$
90° elbow long radius	0.18	Sudden	$h = \dfrac{1}{2}\left[1 - \left(\dfrac{D_1}{D_2}\right)^2\right] V_2^2/2g$
45° elbow	0.18		

FIGURE 7.1 Recommended energy loss coefficient, k, for flanged pipe fittings.

Using the equivalent length approach, the head losses computed for form losses in piping components using Equation (7.6) may be directly added to the head loss due to the friction in the pipe using Equation (7.4) to determine the total head loss through a run of cooling water piping.

TABLE 7.2
Recommended Energy Loss Coefficient, *k*, for Valves Fully Open

Valve type	K
Angle	1.8–2.9
Ball	0.04
Butterfly	
25-lb Class	0.16
75-lb Class	0.27
150-lb Class	0.35
Check valves	
Center-guided globe style	2.6
Double door	
8 in or smaller	2.5
10 to 16 in	1.2
Foot	
Hinged disc	1–1.4
Poppet	5–1.4
Rubber flapper	
$V < 6$ ft/s	2.0
$V > 6$ ft/s	1.1
Slanting disc[d]	0.25–2.0
Swing[d]	0.6–2.2
Cone	0.04
Diaphragm or pinch	0.2–0.75
Gate	
Double disc	0.1–0.2
Resilient seat	0.3
Globe	4.0–6.0
Knife gate	
Metal seat	0.2
Resilient seat	0.3
Plug	
Lubricated	0.5–1.0
Eccentric	
Rectangular (80%) opening	1.0
Full bore opening	0.5

7.3 THE BOWMAN-BAIN EQUATIONS FOR FRICTION LOSS

During preoperational testing of the EECW System at the BFNP during the summer of 1976, certain HXs were found to be receiving inadequate cooling water flow due to a buildup of foreign materials on the interior of the CS piping servicing the equipment. As discussed in Chapter 6, a study was undertaken by TVA to determine the pervasiveness of this problem in the TVA system and to develop recommended practices to mitigate its effects in the design of future power plants.[7] The author supervised the CS raw water piping sampling program and analysis conducted by W. S. Bain of TVA that was initiated to determine the extent of the problem in TVA power plants existing at that time.[7,8] The extent of the buildup of corrosion products that were found in the 50 samples that were removed from nine different TVA power

TABLE 7.3
Equivalent Lengths of Straight Pipe for Piping Components

Nominal pipe size, in	0.5	0.75	1	1.25	1.5	2	2.5	3	4	5	6	8	10	12	14	16	18	20	24	30	36	42	48
Gate valve	0.4	0.6	0.7	0.9	1.1	1.4	1.7	2	2.7	3.4	4	5.3	6.7	8	8.8	10	11.3	13	15	19	23	27	31
90° elbow	1.6	2.1	2.6	3.5	4	5.2	6.2	7.7	10	13	15	20	25	30	33	38	42.2	47	57	70	85	100	115
90° long radius elbow	0.8	1.1	1.4	1.8	2.2	2.8	3.3	4.1	5.4	6.7	8.1	11	13	16	18	20	22.5	25	30	37	45	53	61
45° std elbow	0.8	1.1	1.4	1.8	2.2	2.8	3.3	4.1	5.4	6.7	8.1	11	13	16	18	20	22.5	25	30	37	45	53	61
Std tee thru flow	1	1.4	1.8	2.3	2.7	3.5	4.1	5.1	6.7	8.4	10	13	17	20	22	25	28.1	31	38	47	57	67	77
Std tee branch flow	3.1	4.1	5.3	6.9	8.1	10	12	15	20	25	30	40	50	60	66	75	84.4	94	113	140	170	200	230
Close return bend	2.6	3.4	4.4	5.8	6.7	8.6	10	13	17	21	25	33	42	50	55	63	70.3	78	94	117	142	167	192
Swing check valve	5.2	6.9	8.7	12	13	17	21	26	34	42	51	33	42	50	55	63	70.3	78	94				
Angle valve	7.8	10	13	17	20	26	31	38	50	63	76	100	125	149	164	188	210	235	283				
Globe valve	18	23	39	39	46	59	70	87	114	143	172	226	284	338	372	425	478	533	641				
Butterfly valve						7.8	9.3	12	15	19	23	30	29	35	38	31	35.2	39	47				
90° welding elbow r/d=1						3.5	4.1	5.1	6.7	8.4	10	13	17	20	22	25	28.1	31	38	47	57	67	77
90° welding elbow r/d=2						2.1	2.5	3.1	4	5.1	6.1	8	10	12	13	15	16.9	19	28	28	34	40	46
Miter bend 45°						2.6	3.1	3.8	5	6.3	7.6	10	13	16	16	19	21.1	24	35	35	43	50	58
Miter bend 45°						10	12	15	20	25	30	40	50	66	66	75	84.4	94	14	14	17	20	23

plants as measured by the average reduction in pipe diameter is illustrated in Figure 6.4. Flow tests were performed at the following three fossil power plants: Widows Creek, Kingston, and Gallatin. The objective of these flow tests was to evaluate the effects of corrosion product buildup on pressure drop and to arrive at a means of predicting the friction loss in CS cooling water piping. The sites were selected to cover a range of age as well as a variety of water sources. All tests were made on straight lengths of pipe to avoid consideration of bends. Tees were included in some of the piping systems tested but the pressure drop across the tee was neglected, since the run of the tee was always in line with the test flow and the lateral branch was always closed. An orifice inserted in a length of new piping was installed in each of the piping systems to measure flow rate. The orifice was installed adjacent to the sections of piping where pressure drop measurements were taken. Taps were installed in the lines to allow pressure drop measurements to be made. Mercury manometers were used to measure the pressure drops across the orifice and each section of piping. Samples removed from each test line were analyzed to determine the percent volume reduction of the pipe interior due to the corrosion product buildup.

Samples removed from the 3-inch (7.6 cm) test line at Widows Creek had a substantial amount of iron oxide and silicon oxide buildup, whereas the samples removed from the 6-inch (15.2 cm) line at Kingston were found to have only a small amount of uniform buildup but had very large randomly spaced tubercles [some approaching 2 inches (5.1 cm) in height]. The samples from Gallatin were found to have a more uniform buildup than the Kingston 6-inch (15.2 cm) line but also had large, randomly spaced tubercles. Table 7.4 shows the average diameter reduction of each set of samples.

The corresponding diameter reduction for each test line was then used with the pressure drop test data to develop appropriate equations for predicting pressure drop. The Hazen-Williams and Darcy-Weisbach equations for pressure drop were both considered, with the set of data being treated separately and then analyzed to establish a correlation to the other sets of data. Predictive equations were then formulated to predict pressure drop in CS raw water piping after 40 years of service. The analysis was based on the head loss per 100 feet of pipe so that Equation (7.4) becomes

$$h_{L,f}/100 = 0.2083\ L\left(\frac{100}{C}\right)\left(\frac{Q^{1.85}}{d^{4.8655}}\right) \tag{7.7}$$

TABLE 7.4
Diameter Reduction in Flow Test Samples

Plant	Average Measured Diameter Reduction (in)
Widows Creek	0.405
Kingston	0.133
Gallatin	
Section A	0.270
Section B	0.320

where

$h_{Lf/100}$ = friction head loss in feet per 100 ft of pipe.

Equation (7.8) is a least squares curve fit of the form of Equation (7.7).

$$h_{L,f}/100 = a_1\, Q^{1.85} \tag{7.8}$$

where a_1 is a constant obtained from each set of test data. By setting Equation (7.7) equal to Equation (7.8), one may solve for d as shown in Equation (7.9).

$$d = \left[\left(0.2083/a_1\right)\left(\frac{100}{C}\right)^{1.85}\right]^{\frac{1}{4.8655}} \tag{7.9}$$

For each set of data with a unique value of a_1 a tabulation of values for d that satisfy Equation (7.8) was created.

Per Figure 6.1, the fact that the flow through a CS cooling water pipe occluded by corrosion products is a function of more than the roughness of the interior of the pipe is obvious. Accordingly, a dimensionless parameter, $d*$, defined in Equation (7.10), was found to be useful in correlating the values of d calculated from Equation (7.9) with the measured values of diameter reduction. (See Section 6.2.1.)

$$d* = \frac{\left(d_{nom} - d_{CALC}\right)}{\Delta d_{MEAS}} \tag{7.10}$$

where

d_{NOM} = nominal inside diameter of new pipe
d_{CALCC} = calculated inside diameter of pipe using Equation (7.9)
Δd_{MEAS} = diameter reduction corresponding to the percent volume reduction.
Figure 7.2 shows $d*$ plotted as a function of "C" for all of the pressure drop tests.

The smallest variation of $d*$ occurs at a value of "C" of approximately 57 at a value of $d*$ approximately equal to 2. Equation 7.11 shows the Hazen-Williams equation modified by Bowman and Bain as discussed herein based on a slightly more conservative value of "C" equal to 55 and a diameter reduction equal to twice that measured.

$$h_{L,f}/100 = \frac{0.63\, Q^{1.85}}{\left(d_{NOM} - 2\,\Delta d_{MEAS}\right)^{4.8655}} \tag{7.11}$$

The Darcy-Weisbach equation can be written (for a pipe length of 100 ft) in the form shown in Equation (7.12).

$$h_{L,f}/100 = 3.11f\frac{Q^2}{d^5} \tag{7.12}$$

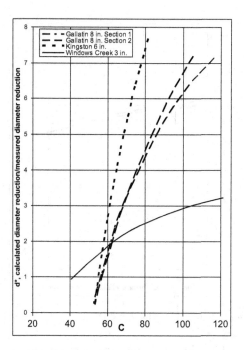

FIGURE 7.2 $d*$ as a function of Hazen-Williams "C" factor.

Equation (7.13) is a least squares curve fit of the form of Equation (7.12).

$$\frac{h_{L,f}}{100} = a_2\,Q^2 \tag{7.13}$$

where a_2 is a constant obtained from each set of test data.

Setting Equation 7.12 equal to Equation 7.13 and solving for the friction factor f yields Equation (7.14).

$$f = \frac{a_2\,d^5}{3.11} \tag{7.14}$$

Equation (7.15) is the expression for the friction factor in fully rough flow by Moody.[9] Full rough flow is almost certain to exist at design flow in old, corroded piping.

$$\frac{1}{\sqrt{f}} = 2\,\log_{10}\frac{3.7}{e/d} \tag{7.15}$$

where
   "e" is in inches.

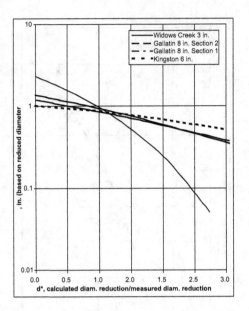

FIGURE 7.3 $d*$ Darcy-Weisbach "e" factor as a function of $d*$.

Equation (7.16) is Equation (7.15) rearranged to solve for e/d.

$$\frac{e}{d} = \frac{3.7}{(10)^{\frac{1}{2\sqrt{f}}}} \tag{7.16}$$

For each pressure drop test with its unique value of a_2, values of f were calculated for different assumed d using Equation (7.14). Equation (7.16) was used to calculate e/d. Figure 7.3 shows a plot of "e" vs $d*$.

The smallest variation in "e" appears to occur at a value of d* equal to 1.0 to 1.2 where $e = 0.9$ to 0.8 inches.

Equation (7.17) shows the Darcy-Weisbach equation as modified by Bowman and Bain using a value of $e = 0.9$ inch and a calculated diameter reduction equal to the measured value of diameter reduction ($d* = 1$).

$$\frac{h_{L,f}}{100} = \frac{3.11 \, f \, Q^2}{\left(d_{NOM} - \Delta d_{MEAS}\right)^5} \tag{7.17}$$

where

$$f = \left\langle 2\log_{10}\left[(4.1)\left(d_{NOM} - \Delta d_{MEAS}\right)\right]^{-2} \right\rangle \tag{7.18}$$

As one may see from Table 7.1, highly regarded sources agree on the adoption of a Hazen-Williams "C" of 100 as the commonly used value for the design of CS piping

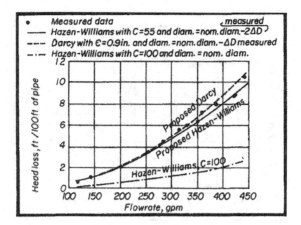

FIGURE 7.4 Comparison of measured and predicted pressure drops for 6-inch pipe at Kingston Fossil Power Plant.

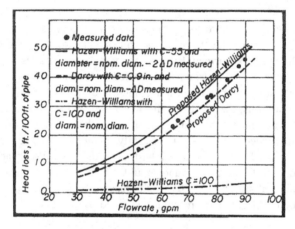

FIGURE 7.5 Comparison of measured and predicted pressure drops for 3-inch pipe at Widows Creek Fossil Power Plant.

in cooling water systems. Such was the case at TVA until it was proven otherwise. Figures 7.4 through 7.6 show the raw data taken for various pressure drop tests along with curves representing different methods of calculating pressure drop using the Bowman–Bain equation.

These curves based upon the empirical evidence contained herein, illustrate the inadequacy of an assumption of C = 100 for design purposes. The Bowman-Bain Equations (7.11) and (7.17), account for not only the roughness of the inside surface of the pipe but also the diameter reduction to predict pressure drop in piping.

At reasonable velocities, the Hazen-Williams equation as modified by Bowman and Bain consistently predicts values of pressure drop greater than the modified Darcy equation, perhaps because the large diameter reduction projected after 40 years is doubled in Equation (7.11) but not doubled in Equation (7.17). As indicated in

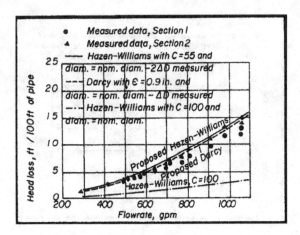

FIGURE 7.6 Comparison of measured and predicted pressure drops for 8-inch pipe at Gallatin Fossil Power.

Figures 7.2 through 7.4, the resulting head loss comparison is based on C = 55 and Δd = 0.4 inch (1.02 cm), since those are the values that resulted in a good correlation for the three flow tests performed. Results may vary for other sites.

Like many power and industrial plants located on the interior, the Tennessee River enjoys relatively soft water having an average pH of 7.3. For those plant sites that have similar conditions, the results of this study should have a profound impact on the design of future raw water piping systems, since the pressure drops calculated by methods recommended herein are significantly greater than those which would be calculated by presently recognized standard methods. It should be noted that these results apply only to friction losses and should not be applied to form losses.

In 1986, the results of the TVA flow and pressure drop tests were presented at an EPRI workshop on MIC.[10] The presentation gained little notice.

7.4   FLOW AND PRESSURE DROP TESTS OF LARGE DIAMETER PIPING

The system friction losses from one point (1) to another point (2) in the system may be arrived at by first calculating the form losses in the system and subtracting the form losses from the dynamic head loss calculated from Equation (7.19).

$$h_L = \left(P_1 - P_2 \right) + \left(Z_1 - Z_2 \right) + \left(VH_1 - VH_2 \right) \tag{7.19}$$

where
h_L = total head loss
P_1 and P_2 = gauge pressure
Z_1 and Z_2 = elevations of the gauge pressure
VH_1 and VH_2 = velocity head.

The velocity head is calculated from Equation (7.20).

$$VH = \frac{V_{pipe}^2}{2g} \tag{7.20}$$

where

V_{pipe} = velocity in the pipe

Form losses, $h_{L,F}$, are calculated based on the equivalent lengths and subtracted from the total head loss to yield the friction head loss, $h_{L,f}$. (See Section 7.1.)

At TVA's SNP ERCW system flow tests were conducted annually from 1980 to 1993.[11] These test data yielded information to calculate the apparent "C" factor in two large headers between the IPS and the auxiliary building. Each of the two 4000-foot (1,220 m) runs of piping includes 24-inch (61.0 cm), 30-inch (76.2 cm), and 36-inch (91.4 cm) diameter sections of piping in series with varying flow rates in the different sections of piping.

For a run of piping with multiple line sizes in series, the Hazen-Williams equation as modified by Bowman and Bain may be expressed as Equation (7.21).

$$\frac{h_{L,f}}{100} = 0.2083 \left(\frac{100}{C} \right)^{1.85} \sum_{n=1}^{s} \left[\left(\frac{Q_n^{1.85}}{\left(d_{NOM,n} - 2\,\Delta d_{MEAS} \right)^{4.8655}} \right) \left(\frac{L_n}{100} \right) \right] \tag{7.21}$$

where

s = number of pipe sections
Q_n = flow in each section of pipe, gal/min
$d_{NOM,\,n}$ = nominal pipe diameter in each section of pipe, in
L = length of pipe in each section, ft.

Solving for "C" yields

$$C = \left\langle \left(\frac{0.2083 \times 100^{1.85}}{h_{L,f}/100} \right) \sum_{1}^{s} \left[\left(\frac{Q_n^{1.85}}{\left(d_{NOM,n} - 2\,\Delta d_{MEAS} \right)^{4.8655}} \right) \left(\frac{L_n}{100} \right) \right] \right\rangle^{\frac{1}{1.85}} \tag{7.22}$$

At TVA's Sequoyah plant, flow tests were conducted on the ERCW system by measuring the cooling water pressures at the IPS and at the auxiliary building. The Sequoyah test data have yielded sufficient information to calculate the apparent "C" factor in the two large headers between the IPS and the auxiliary building. Each of the two runs of piping has varying flow rates in the different sections of piping.

Since it is impractical to remove a section of large-diameter piping to take a measurement, a value of $\Delta d_{MEAS} = 0.2$ in. (0.51 cm) was assumed based on Figure 6.4 and the age of the piping. However, the "C" factor calculated from Equation (7.22) is relatively insensitive to the value of Δd_{MEAS} for such large diameter pipe.

Figures 7.7 and 7.8 show the results of these tests for headers 2A and 2B, respectively.

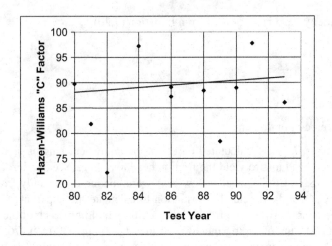

FIGURE 7.7   Sequoyah Nuclear Plant ERCW header 2A Hazen-Williams "C" factor.

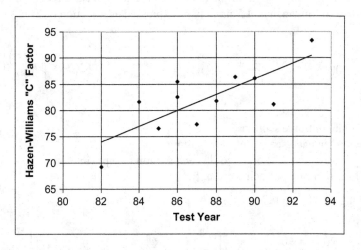

FIGURE 7.8   Sequoyah Nuclear Plant ERCW header 2B Hazen-Williams "C" factor.

The results are somewhat scattered even though these tests were performed with calibrated instruments and the data was independently verified. However, this scatter is to be expected, since the tests involved measuring pressures with separate pressure gauges approximately one-half mile apart and measuring flows in 30-inch (76.2 cm) piping with standard flow orifices.

The results are somewhat surprising, because the "C" factors do not appear to be decreasing with age as expected. However, as discussed in Section 6.5, the piping was first placed in service in 1977 and 1978. By 1982, the "C" factor had degraded to approximately 90 for Header A and 70 for Header B when the biocide treatment was implemented. The fact that Train A is normally operated, while Train B is not, may be a consideration. (See Figure 6.13.)

TABLE 7.5

Watts Bar Nuclear Plant ERCW Header Hazen-Williams "C" Factors Before and After Lining with Cement Mortar In Situ

Supply Header	"C" Factor for CS	"C" Factor for CML
1A	74.6	163.7
2A	80.7	184.8
1B	77.1	219.5
2B	75.4	154.1
Average	77.0	181.5

In 1982, TV A conducted similar flow test on the supply headers at the WBNP before and after the CML was applied as discussed in Section 5.4.5.[12] The Hazen-Williams "C" factor for the piping before and after the CML was applied is shown in Table 7.5.

The increase in flow-passing capability of the piping is quite dramatic. The result has been an increase header pressure in the ERCW system inside the auxiliary building. The average value for CML is even higher than the published Values of "C" = 130-140. (See Table 7.1.) The scatter in the results of the CML tests when compared to the CS tests is due to the lower head loss measured relative to the test instrument accuracy. These tests were conducted prior to completion and commercial operation of WBNP, so each header was operated about the same amount of time and without biocide treatment.

In 1983, a similar flow and head loss test was conducted on the BFNP EECW system. The EECW system pressure was measured at the EECW pump discharge and at the entrance to the emergency diesel generators. The flow to the diesel generator HXs was measured in the 14-inch (35.6 cm) and 18-inch (45.7 cm) CS pipes from the pumps. The "C" factor calculated by the procedure described in this section was C = 88 assuming $\Delta d = 0.2$ (0.51 cm) in.

7.5 MULTIFLOW PIPING ANALYSIS SOFTWARE

Realizing that the cooling water piping systems engineered prior to 1980 were inadequate because they were designed on the basis of Hazen-Williams C = 100, TVA initiated a reanalysis of the safety-related cooling water systems in all of their nuclear plants. Since this analysis was highly labor intensive, software (MULTIFLOW) was developed for hydraulic analyses of raw water and condensate piping networks in nuclear power plants by the TVA Norris Engineering Laboratory.[13] MULTIFLOW incorporates TVA's established hydraulic design standards[6] into an intuitive, menu-based, computer code for balancing steady, incompressible flows and temperatures in raw water, and condensate pipe networks of arbitrary complexity and provides output reports.

A network is built by specifying links, nodes, tees, and boundary conditions using a series of menus and dialog boxes. Extensive built-in tabular data based on TVA hydraulic design standards provide appropriate default values for internal diameters and roughnesses of piping, equivalent lengths of various fittings, and loss coefficients for combining and dividing flow through tees. Available inputs include global or local specifications of corrosion effects on pipe roughness and inner diameter values. The network input data are saved in text format to a disk file that may be edited later using either the editing capabilities available in MULTIFLOW or any text editor. After all necessary input data are provided, MULTIFLOW solves the system and provides output reports to a printer or to the monitor.

MULTIFLOW was developed and tested in accordance with the requirements of TVA Nuclear QA criteria for computer software. The requirements included validation by comparing results obtained using MULTIFLOW with results obtained by hand calculations for an extensive set of problems exercising all of the capabilities of MULTIFLOW. In addition to its adherence to TVA QA criteria and mechanical design standards, MULTIFLOW has several input, solution, and output capabilities that are not usually available in general-purpose network codes. The size of a network solved using MULTIFLOW is limited only by available computer memory. MULTIFLOW iteratively computes flows through tees using separate loss coefficients for each flow path that depend on the ratio of flow through the path to total flow through the tee. MULTIFLOW computes fluid temperatures throughout a network based on specified supply flow temperatures and specified loads at HXs. For condensate systems, fluid densities, viscosities, and vapor pressures are treated as variable with temperature and pressure. MULTIFLOW allows the flow in a link to be fixed at a user-specified value and computes the valve flow coefficient or pump total head required to achieve the desired flow. Components handled by MULTIFLOW include pipes, pumps, check valves, pressure regulating valves, relief valves, tees, fixed flows, and various sources of energy loss (e.g., valves, elbows, reducers, etc.). MULTIFLOW also contains built-in design diagnostic tools that flag unacceptable conditions, such as pressures below the vapor pressure.

7.6   PIPING NETWORK ANALYSIS USING MULTIFLOW

As discussed in Section 3.3 and 10.1, Section 316(b) of the CWA requires power and industrial plants that draw water from the waters of the United States to implement the BTA to reduce injury or death of fish and other aquatic life. The first alternative proposed by the EPA to accomplish the stated goal was to install a closed-cycle recirculating heat rejection system.

In the 1970s to comply with the CWA, TVA constructed two NDCT at the SNP to operate in a "helper" mode to cool the CCW prior to discharging it into the Tennessee River. Provision was made in the modification to be able to operate in closed-cycle recirculating mode by means of a channel back to the CCW pumping station. Provision was also made to add a third NDCT if required.

For many years, compliance with Section 316(b) of the CWA was subject to litigation. Final rules for compliance with this section were promulgated by the EPA in 2014. In 2017, TVA conducted a study at the SNP to determine the best method of complying with Section 316(b) of the Act.[14] Section 3.2.1.3 of Reference 14 describes Alternative A, which is to operate the existing NDCT in closed-cycle recirculating mode as follows:

- Alternative A – Refurbish the existing two NDCTs

Alternative A would require the refurbishment of the existing two NDCTs including the removal, disposal, and replacement of the existing fill material, hangers, and drift eliminators. The fact that no additional heat dissipation capability is provided for Alternative A results in the maximum temperature entering the condenser and RCW System being as much as 10°F higher than is presently the case and approximately 4°F higher than is the case for Alternatives B and C. To compensate for the higher temperature of the cooling water entering the RCW System, the flow rate to the HXs served by the system must be increased, and not only must much of the CS RCW piping be replaced with SS which exhibits superior flow-passing capability, but also the existing RCW Pumps would have to be replaced with higher head pumps. Overall, Alternative A results in the most significant loss at 55 MWe per Unit of SQN's generation output or a reduction of 4.5% per unit per year.

The RCW system for each nuclear unit at SNP takes its suction from the CCW supply conduit upstream of the MC. There are four 20" (50.8 cm) RCW strainers and five horizontal, centrifugal, dry-pit RCW pumps. One of the RCW pumps is an installed spare. The RCW system serves innumerable non-safety-related HXs in the SNP such as the turbine lube oil cooler, stator water cooler, hydrogen cooler, and a host of smaller HXs serviced by small CS piping. The RCW system discharges back into CCW discharge conduit at various points downstream of the MC. MULTIFLOW was utilized to determine which CS RCW piping would be required to be replaced with SS.

Measurements of the RCW system pressures were made at the RCW pump suction and discharge and at 15 points in the system. Table 7.6 show the results of MULTIFLOW runs for an array of Hazen-Williams "C" factors and values of Δd.

The results of the test implied that the Hazen-Williams "C" factor and the diameter reduction that best modeled the actual pressures in the SNP RCW system are C = 80 and Δd =0.15 inch (0.381 cm), with an average difference between the measured pressure and the pressure calculated using MULTIFLOW of only 0.3 psi (2.1 kPa).

TABLE 7.6

Sequoyah Nuclear Plant RCW System Pressure Difference* For an Array of Hazen-Williams "C" Factors and Δd

Hazen-Williams		Average
"C" Factor	Δd	ΔP (psi)
100	0	7.6
90	0.3	−1.0
85	0.2	−1.8
85	0.3	−0.5
80	**0.15**	**0.3**
80	0.2	0.5
80	0.3	0.3
70	0.3	2.3

* ΔP = Measured pressure – calculated pressure w/MULTIFLOW.

TABLE 7.7
Sequoyah Nuclear Plant RCW System Piping Replacement

Line size (in)	Length (ft)	Elbows	Gate	Globe	Reducer	Butterfly
0.75	44	8	8	4	1	
1.00	994	44	15	13	17	
1.50	229	20	5	3	2	
2.00	1803	58	14	18	4	
2.50	573	68	20	1	12	
3.00	471	25	3	6	4	
4.00	379	26	2	1		2
6.00	76	7			1	
14.00	104	4				

Table 7.7 shows which CS RCW piping would need to be replaced with SS. The RCW Pumps would also need to be replaced with higher head pumps.

REFERENCES

1. Linsley, R. K. et al. *Water-Resources Engineering*, 4th ed., pp. 347–349, McGraw-Hill, Inc., 1992.
2. Hansen, E. G. *Hydronic System Design and Operation-A Guide to Heating and Cooling with Water*, p. 61, 63, McGraw-Hill, Inc., 1985.
3. *Crane Technical Paper No. 410*, 24th Printing, Crane Co., King of Prussia, PA, 1988.
4. *Cameron Hydraulic Data*, 16th ed., pp. 3–7, 3–8, 3–20, Ingersoll-Rand, Woodcliff Lake, NJ, 1981.
5. Mays, L. W., *Water Resource Engineering*, 2nd ed., pp. 473, 489–492, John Wiley & Sons, Inc., 2011.
6. *Mechanical Design Standard DS-M3.5.1, Rev. 7, Pressure Drop Calculations for Raw Water Piping and Fittings*, Tennessee Valley Authority, Knoxville, TN, 1999.
7. Bain, W. S., et al. *Corrosion in Carbon Steel Raw Water Piping*, (TVA Ref. ESS800416204) Tennessee Valley Authority, Knoxville, TN, 1979.
8. Bowman, C. F. and W. S. Bain. A New Look at Design of Raw Water Piping. *Power Engineering Magazine*, pp. 73–77, Vol. 48, No. 8, August 1980.
9. Moody, L. F. Friction Factors for Pipe Flow. *Transactions of the ASME*, Vol. 66, No. 8, pp. 671–684, 1966.
10. Bowman, C. F. *Microbe – Induced Corrosion of TVA Raw Cooling Water Piping*. *Proceedings of the EPRI Workshop on Microbe Induced Corrosion (MIC)*, Electric Power Research Institute, 1986.
11. Bowman, C. F. Solving Raw Water Piping Corrosion Problems. *Power Engineering Magazine*, Vol. 98, No. 7, pp. 35–38, July 1994.
12. Bowman, C. F. *In Situ Cement-Mortar Lining of Safety-Related Service Water Piping Systems*. International Joint Power Conference, 94-JPGC-NE-6, 1994.
13. Schohl, G. A., et al. *MULTIFLOW, a Quality Assured Intuitive Computer Code for Hydraulic Analysis of Pipe Networks*. Proceedings of the 1995 International Joint Power Generation Conference, Vol. 2, pp. 67–79, 1995.
14. Sequoyah Nuclear Plant §316(b) – §122.21(r)(10) – (13) Information, Appendix A to Transmittal from TVA to Tennessee Department of Environment and Conservation, June 29, 2018.

8 Cooling Water System Engineering Pitfalls

8.1 HYDRAULIC PRESSURE GRADIENT

The hydraulic pressure gradient shown in Figure 8.1 illustrates the importance of maintaining a positive gauge pressure throughout the cooling water system. The head loss indicated by the hydraulic gradient is calculated as discussed in Chapter 7.

Consider a cooling water pump with pressure at the discharge of the pump sufficient to pump the water 140 feet (42.7 m) above the lake level, which is at zero gauge pressure. The hydraulic pressure gradient represents the head loss through the cooling water system. At any point in the system, the gauge pressure is the difference between the hydraulic gradient and the elevation of the piping or component in the system where the pressure is measured in lbf/in² (or kPag). In a cooling water system, there are likely to be changes in elevation. To avoid serious problems, one should ensure that the hydraulic gradient does not fall below zero at the highest elevation points. This is especially true for a HX, which is often located relatively high in a power or industrial plant as illustrated by Figure 8.2. If the gauge pressure drops at or below zero lbf/in² (kPag), depending on the temperature of the cooling water, vaporization, and/or air liberation can occur as discussed below. One solution to this problem is to increase the head of the cooling water pump and locate a throttling

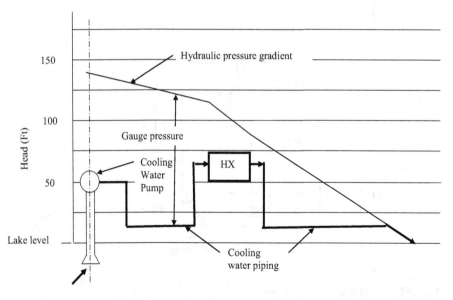

FIGURE 8.1 Hydraulic gradient with positive gauge pressure.

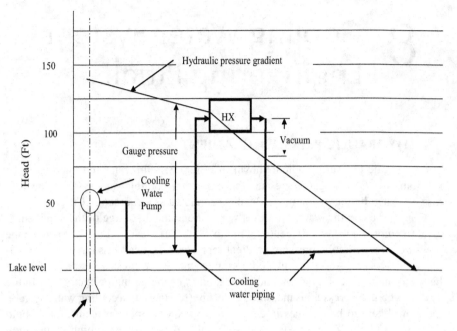

FIGURE 8.2 Hydraulic gradient with negative gauge pressure.

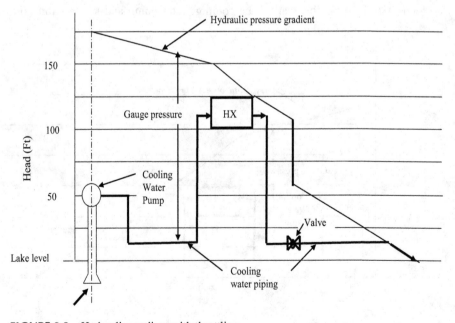

FIGURE 8.3 Hydraulic gradient with throttling.

valve downstream of the HX to provide a backpressure in the HX as shown in Figure 8.3. Another solution would be to locate the HX lower in the plant.

The information in Sections 8.2 through 8.4 is largely taken from Reference 1 by F. P. Carr.

8.2 AIR LIBERATION

Since the atmospheric gases (primarily nitrogen and oxygen) are soluble in water, significant quantities of these gases can be absorbed and carried into the piping system. These gases can also cause flow disturbances and reduced heat transfer rates in HXs if they come out of solution inside the system.

Henry's law relates the important variables affecting gas solubility as shown in Equation (8.1).

$$N_x = \frac{P_x}{K_x} \tag{8.1}$$

where
N_x = mole fraction of dissolved gas, x
P_x = partial pressure of dissolved gas x
K_x = Henry's Law coefficient for gas x.

This equation shows that the number of moles of gas that can be dissolved in a given quantity of water is directly proportional to its partial pressure and inversely proportional to its Henry's law coefficient. For a particular gas, the latter coefficient is primarily a function of temperature, increasing as the temperature increases. Thus, water becomes less soluble to air as the pressure decreases or the temperature increases. Since this is exactly what happens as cooling water passes through the cooling water system, it is possible for gases to be released within the system.

Air liberation within the system can affect heat transfer of system HXs. For example, if a HX were to operate with a sufficiently high temperature rise, or was located so that the local pressure was sufficiently low, air liberation could occur. This would not necessarily create any problems. However, if the geometry of the HX or the cooling water piping were such that liberated air could be trapped within the HX, the air could displace the water and effectively reduce the available heat transfer area. This could occur in HXs with a bottom outlet or where the cooling water piping turned down at the outlet nozzle.

Air liberation can also affect fluid flow if conditions are such that the flow is in the so-called "slug" flow regime. In this regime, large bubbles of liberated gases are periodically swept through the system, so the cooling water flow is not steady but pulsing and unacceptable vibration can also result.

The preferred approach in initial design is to avoid air liberation entirely. The first step to achieve this goal is to calculate the maximum quantity of air dissolved in the cooling water inlet. Henry's law may be used in the following combined form, which includes the effect from all atmospheric gases as shown in Equation (8.2).

$$N_{air} = \frac{P_b}{K_{air}} \tag{8.2}$$

where

N_{air} = mole fraction of all dissolved gases

K_{air} = Henry's Law coefficient from Figure 8.4.

The barometric pressure should be the standard atmospheric pressure corresponding to the elevation of the free surface water level. To maximize N_{air}, the Henry's law coefficient, K_{air}, should be based on minimum cooling water supply temperature.

The final factor, C, accounts for the degree of saturation of the inlet water. Since it is common for bodies of water to approach 100 percent saturation in the winter, "C" should not be less than 1.0. However, it is also known that the water can be supersaturated with air downstream of dams or where the source of cooling water is an evaporative cooling device such as a cooling tower or spray pond. (See Chapter 10.) Evaporative cooling devices may therefore justify a "C" value of 1.5, whereas a cooling lake may dictate a value of 1.0. Hydroelectric turbine operation can cause saturation up to 100 percent, while spillways and evaporative cooling devices can cause an increase to 150 percent. Having determined the mole fraction of dissolved air in the inlet cooling water, the next step is to determine the minimum pressure required to keep this air in solution. Again, use Henry's law as shown in Equation (8.3).

$$P_{required} = \frac{N_{air-max}}{K'_{air}} \tag{8.3}$$

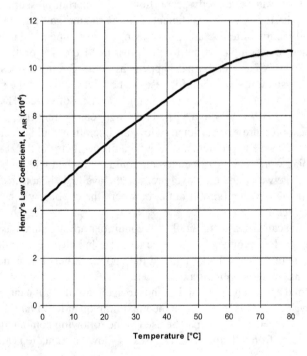

FIGURE 8.4   Henry's law coefficient for air.

where

$P_{required}$ = minimum gauge pressure to keep air in solution
$N_{air\text{-}max}$ = maximum mole fraction of air in the entering cooling water
K'_{air} = Henry's Law coefficient based on the HX outlet temperature from Figure 8.4.

The key difference is that K'_{air} is based on the HX outlet temperature. The outlet temperature may be calculated using Equation (8.4).

$$T_{out} = T_{in} + \frac{Q}{m \; c_p} \qquad (8.4)$$

where

T_{out} = HX outlet temperature
T_{in} = HX inlet temperature
Q = maximum HX heat load
m = cooling water mass flow rate
c_p = constant pressure specific heat.

Since T_{out} is greater than T_{in}, K'_{air} in Equation (8.3) will be greater than K_{air} in Equation (8.2) as evident from Figure 8.4. T_{out} must be calculated for each HX and $P_{required}$ be determined for each different value.

The next step is to determine the actual minimum gauge pressure, P_{actual}, on the discharge side of each HX. This involves simply calculating the hydraulic gradient as discussed in Section 8.1. If "P_{actual}" is less than "$P_{rerquired}$", the potential exists for air to be liberated within the HX and/or downstream piping due to the increase in temperature. The fact that air is released is not necessarily unacceptable. The flow regime must be determined to make the evaluation of those having the potential for air liberation by first calculating the Froude number with Equation (8.5) and then consulting Figure 8.5.

$$Fr = \frac{V^2}{gD} \qquad (8.5)$$

where

Fr = Froude number
V = local velocity
g = acceleration due to gravity
D = pipe or tube inside diameter.

Next one must calculate the air-to-water volume ratio, Vr, of air liberated at the pipe or tube section being investigated. To determine this ratio, one must first find the actual mole fraction of air released. This will be the difference between the mole fraction of dissolved air in the inlet water, N_{air}, and the mole fraction that local conditions will permit to remain in solution, "N'_{air}" shown in Equation (8.6)

$$N'_{air} = \frac{P'}{P_b K'_{air}} \qquad (8.6)$$

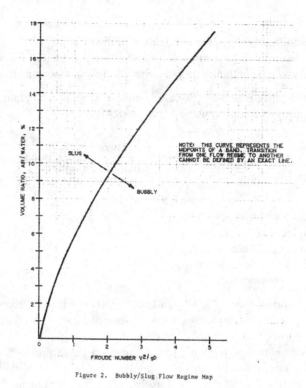

Figure 2. Bubbly/Slug Flow Regime Map

FIGURE 8.5   Bubbly/slug flow regime map.

where
N'_{air} = mole fraction that local conditions will permit to remain in solution
P' = local static pressure
K'_{air} = Henry's Law coefficient at local temperature (from Figure 8.4).

The volume of air liberated per unit volume of water, Ve, is then calculated by first considering the ideal gas law shown in Equation (8.7) and (8.8).

$$PV = nRT \qquad (8.7)$$

$$V = \frac{nRT}{P} \qquad (8.8)$$

where
P = pressure
V = volume
n = number of moles
R = universal gas constant
T = absolute temperature

The volume of air liberated per unit volume of water, V_r, may then be calculated by Equation (8.9).

$$V_r = \frac{\left(N_{air} - N'_{air}\right)RT'}{P'} \tag{8.9}$$

where

V_r = air-to-water volume ratio

T' = absolute local temperature.

The values of "Fr" and "V" are then plotted on Figure 8.5. Those that fall within the bubbly flow regime are generally acceptable. Those in the slug flow region may require modification.

Many possibilities exist to correct HX loops found to be in the slug flow regime. However, many of the solutions become increasingly difficult as design and construction progress. Perhaps the simplest solution is to relocate the flow-balancing valve beyond the region having the potential for air liberation as shown in Figure 8.3. This will pressurize the critical area by the amount of the differential across the valve, possibly eliminating the potential for air release or transforming it to bubbly flow. Another possibility is to reroute the piping to eliminate problem high points. It may even be possible to relocate a HX to a lower floor as shown in Figure 8.1. Booster pumps to pressurize problem areas are also effective, particularly on the supply side of the HXs.

8.3   FLASHING

The term "flashing" refers to the change of state from liquid to gas. It is generally associated with the formation of large volumes of vapor as a result of heating, expansion into a low pressure region, or both. Flashing might occur, for example, within a HX, because of the temperature rise of the water. The effective heat transfer area would thereby be reduced, leading to a further rise in temperature, more flashing, etc. Eventually the HX could become completely filled with vapor and unable to remove its design heat load.

Flashing can also occur in regions of low pressure. Such regions might exist downstream of flow-balancing valves or automatic temperature control valves. For this reason, such valves should generally be located downstream of their associated HXs to preclude the possibility of collecting vapor within the HX.

The first step in evaluating the potential for flashing is to determine the HX outlet temperature as shown in Equation (8.4). The saturation vapor pressure, P_{sat}, corresponding to "T_{out}" may then be read from the *ASME Steam Tables*.[2,3] Next, the local HX outlet pressure is calculated as described in Section 8.1. If $P_{local} < P_{sat}$, as illustrated in Section 8.2, flashing will occur. The outlet of flow-balancing valves or temperature control valves should also be checked for flashing. Flashing should not be allowed to occur within system HXs. Possible corrective measures include all of those mentioned in Section 8.2 for air liberation. It might also be possible to prevent flashing by increasing the design flow rate of the cooling water, thereby decreasing the temperature rise. HX performance should be evaluated when determining the feasibility of the latter option.

Flashing downstream of throttling valves may lead to unacceptable levels of cavitation, vibration, or surging. Criteria for evaluating this problem are presented in the next section. If found to be unacceptable, and if the local temperature rather than local pressure is the main problem, bypassing a portion of the cooling water inlet flow around the HX to mix with valve inlet flow might effectively prevent flashing.

8.4 CAVITATION

Whereas flashing deals with the formation of vapor, cavitation deals with the damaging effects as the vapor bubbles change back to liquid. The implosion of the vapor cavities can generate extremely high local pressures. The shock wave itself is sufficient to fracture the grain boundaries of metal surfaces, causing removal of whole grains and leaving a roughened metal surface. Erosion may lead to structural failure or leakage. Corrosion accelerates the process.

Cavitation is generally associated with throttling valves, orifice plates, venturis, and other such devices that create a sudden change in local pressure. The propensity for and severity of cavitation may be determined by calculating the "cavitation index" shown in Equation (8.10).

$$K_i = \frac{P_d - P_{sat}'}{P_u - P_d} \tag{8.10}$$

where
   K_i = cavitation index
   P_u = local upstream pressure
   P_d = local downstream pressure
   P_{sat} = saturation vapor pressure at the maximum operating temperature.

The acceptable cavitation level or index, "K_i", so calculated may then be interpreted from Figure 8.6. If unacceptable cavitation is indicated by this procedure, the corrective actions in Sections 8.2 for air liberation and/or 8.3 for flashing may be effective. Generally, the low outlet pressure is the major contributor to the cavitation problem, and steps to increase it will be effective. Adding a multihole restricting orifice downstream of the cavitating component will increase the backpressure of the component and reduce or eliminate the cavitation. However, the effect on system performance at off-design points and over the plant life must be carefully considered. If other measures are not feasible, the use of a special anti-cavitation valve or multiple orifices might be considered. Initial piping design should attempt to minimize the required throttling in the high-flow HX loops.

If the cooling water system design uses vertical pumps, provisions should be made for air and vacuum, release between the pump discharge and the discharge check valve. (See Figure 3.1.) The vacuum relief valve is required when the difference in pump discharge elevation and the design low water level is greater than approximately 33 feet (10.1 m) to prevent column separation when a pump is shut

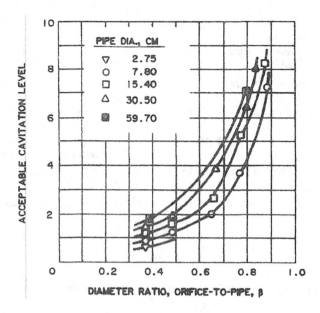

FIGURE 8.6 Cavitation indices for orifices (from Reference 4).

down. When the pump is restarted, the air in the piping must be vented to prevent air from being forced into the cooling water system, which could degrade the performance of HXs served by the system.

8.5 WATER HAMMER

The information in this section is largely taken from Reference 5 by G. C. Dunn and J. D. Hubble.

8.5.1 CONCEPTS OF WATER HAMMER

Water hammer is a hydraulic pressure transient caused by changes in fluid velocity and the resulting conversion of fluid kinetic energy into pressure energy, which is propagated as a pressure wave. Analysis and control of water hammer requires some understanding of this pressure wave propagation. The fundamental equations involved are presented and discussed in the remainder of this section.

The equations of continuity and momentum may be applied to obtain the magnitude of change in fluid pressure as a function of the magnitude of change in fluid velocity.[6] The instantaneous pressure increase is, in general, related to the velocity decrease by Equation (8.11).

$$H - H_o = \frac{A}{g}\left(V_o - V\right) \qquad (8.11)$$

where

H = pressure head
H_o = initial pressure head
A = wave speed
g = gravitational constant
V = fluid velocity
V_o = initial fluid velocity.

Equation (8.11) applies only for a simple wave propagation when the deceleration in fluid velocity from "V_o" to "V" occurs over a period of time, T, where $t < 2L/A$, where "L" is the length of pipe and where the wave speed, A, is much greater than the initial fluid velocity, V_o. The pressure wave thus generated travels through the pipe. The travel time required for the return of this wave is defined in Equation (8.12).

$$T = \frac{2L}{A} \tag{8.12}$$

An expression for wave speed in an infinite reservoir in English units as derived in Reference 6 is shown in Equation (8.13)

$$A = \sqrt{\frac{E_w}{\rho_w}} \tag{8.13}$$

where

E_w = bulk modulus of elasticity of water
ρ_w = density of water.

The wave speed in a pipe depends on the material, wall thickness, and type of construction as well as the fluid properties and temperature. Equation (8.14) (in English units) shows how to compute the wave speed as a function of the particular piping system.[7]

$$A = \sqrt{\frac{E_w}{\rho}} \; x \; \sqrt{\frac{1}{1 + \left(\dfrac{E_w \, D}{E_p \, t}\right)}} \tag{8.14}$$

where

E_p = bulk modulus of elasticity of the pipe material
D = pipe diameter
t = thickness of the pipe wall.

Wave speeds for large steel pipe may be as low as 3,000 ft/sec (914 m/s) for small, high-pressure steel pipe or as low as 4,500 ft/sec (1,372 m/s) in the absence of ingested air.

To calculate the wave speed in a homogeneous mixture of air and water, Equations (8.15) and (8.16) are used along with Equation (8.14).

$$E_M = \frac{E_w}{1 + \dfrac{VM_a}{VM_T}\left(\dfrac{E_w}{E_a} - 1\right)} \qquad (8.15)$$

$$\rho_M = \rho_a \left(\frac{VM_a}{VM_T}\right) + \rho_w \left(\frac{VM_w}{VM_T}\right) \qquad (8.16)$$

where
E_M = bulk modulus of elasticity of the mixture of water and air
E_w = bulk modulus of elasticity of water
E_a = bulk modulus of elasticity of air
VM_w = volume of water
VM_a = volume of air
VM_T = total volume
ρ_M = density of the mixture of water and air
ρ_w = density of water
ρ_a = density of air.

Measurements of wave speeds as low as 1000 ft/sec (305 m/s) with 0.4 percent air (by volume) in water have been published which show good agreement with these equations.

Therefore, the assumption that the wave speed, A, is much greater than the initial fluid velocity, V_o, in Equation (8.11) is valid.

When a change in fluid velocity occurs over a time interval greater than the travel time of the resulting pressure wave as expressed in Equation (8.12), the change is considered slow, and the governing equations in a frictionless system are shown in Equations (8.17) and (8.18).[8]

$$H - H_o = F + f \qquad (8.17)$$

$$V - V_o = -\frac{g}{A}\left(F - f\right) \qquad (8.18)$$

where
F = the magnitude of the pressure wave generated by a change in water velocity and whose direction is opposite to original water flow
f = the magnitude of the reflected pressure wave generated by a change in water velocity and whose direction is that of the original water flow.

where "F" is a function of t-x/A and f is a function of t+x/A and "x" is the distance along the pipe.

Note that "x" is positive in the direction opposite the initial fluid velocity, since it measures the distance upstream from the point of initial disturbance in flow.

Equation (8.17) implies that at a time, t, at a point "x" feet from the initial disturbance, the head rise is equal to the algebraic sum of the pressure wave, F, and the reflected wave, f. These waves are propagated in opposite directions in the pipe with a constant velocity of "A" feet per second. When an "F" wave passes an "f" wave, neither wave is attenuated or undergoes a change in shape.

Equation (8.18) relates the magnitude of the pressure waves "F" and "f" to the change in velocity which propagates the waves.

Water hammer is generally caused by changes in the mode of equipment operation. This, in turn, is most often the result of valve operation, startup, shutdown, or inadequate venting actions.

8.5.2 CONTROL AND ANALYSIS OF WATER HAMMER

Rapid valve operation will cause water hammer as described by Equation (8.11). If valve operation time is less than "T" as defined in Equation (8.12), the generated pressure waves will reflect against the closed valve.

By solving Equation (8.17) for "F" and substituting into Equation (8.18), an expression for the change in pressure head resulting from these conditions may be written as shown in Equation (8.19).

$$ H - H_o = -\frac{A}{g} \left(V - V_o \right) + 2f \qquad (8.19) $$

Initially, "f" equals zero at the valve. It may be shown that for a wave reflected from a reservoir, the first reflected wave at the valve is equal in magnitude but opposite in sign to the direct wave which left the valve "T" seconds earlier.[6]

For slow valve operation, the severity of potential water hammer may be determined by approximating the actual valve closure by a series of successive instantaneous stepwise movements at some time intervals, Δt, where $\Delta t < T$.

In order to determine the changes in velocity for a chosen time interval, a relationship is required between the percent valve opening with respect to time (during valve operation) and the valve coefficient as a function of valve opening. With this information, the change in flow and corresponding change in velocity may be calculated for the specified time interval. Equation (8.19) is then used to calculate $H - H_O$ for the selected interval. For each stepwise movement, the initial values of "H" and "V" are taken as the final value calculated from the previous step. This procedure is repeated until the valve reaches its final position. Given a specified design, this procedure allows the effects of valve operation to be evaluated.

Long pipe lengths will require relatively long valve closure times in order to prevent water hammer. Equation (8.2) reveals that this factor is usually negligible unless there are several thousand feet of pipe. However, consideration must be given to the transient resulting from an instantaneous valve closure due to a motor operator failure. If such a failure would result in an unacceptable impact on the system, snubbers (dampers) should be furnished with the valve operator.

Unless properly designed, a cooling water system is likely to experience water hammer during startup and shut down due to the acceleration and deceleration of fluid columns inherent in these operations. By proper design and operation, water hammer can be avoided by controlling the rate of change of fluid column velocity.

During a cooling water pump start, the acceleration of the fluid column may be controlled by starting the pump against a closed or partially closed discharge valve and then gradually opening the valve. The reverse procedure may be followed when shutting down a pump. Spurious pump trips can result in column separation in the pump discharge, with resulting water hammer. The vacuum generated gradually overcomes the receding water column with a spring-like action, and then collapses as the column of water is reunited. A vacuum breaker/air relief valve may be used on the pump discharge to prevent formation of the temporary vacuum.[9] Check valves or automatic-closing butterfly valves located on pump discharges may also aid in reducing the detrimental effects of column separation at high points in the system by keeping the upstream leg of piping full of water following a pump trip.

High points in the system which would be pressurized under normal operating pressures, will experience depressurization and column separation upon system shutdown if the high point is more than 33 feet (10.1 m) above the hydraulic gradient with the system depressurized. Upon startup, as the system is re-pressurized, the vapor column will collapse. This results in a water hammer as the column of water is reunited. Vacuum breaker/air relief valves should be provided at all such high points to prevent water hammer as the column is reunited. Carefully located check valves in the system will often provide protection against column separation at high points in the system by keeping the upstream leg of piping full of water following a system shutdown.

During startup, the system should be filled carefully to vent all air pockets at high points so that flow will not be impeded. Care must be exercised in the sizing of vent valves, since the size of the vent valve establishes the velocity at which the column of water rises at the high point as the air is displaced by water. If the vent valve is too large, resulting in the water column rising at a relatively high velocity, the resulting deceleration as the last of the air is released can be very significant. This problem can be particularly serious in large tanks and HXs. An alternative to careful sizing of the vent valve would be to place an orifice in the vent line sized to restrict the rate at which the air may be vented.

It should be emphasized that the complexity and potential consequences of water hammer lead to design as the preferred method for control of hydraulic transients. However, a detailed water hammer analysis may be required if the system design parameters violate one or more of the guidelines described herein.

REFERENCES

1. TVA Mechanical Design Guide DG-M6.3.3. *General Design of Essential Raw Cooling Water Systems*, Tennessee Valley Authority, Knoxville, TN, 1986.
2. Meyer, C. A. et al. *ASME Steam Tables*, 6th ed., American Society of Mechanical Engineers, New York, NY, 1993.
3. *ASME Steam Tables Compact Edition*, American Society of Mechanical Engineers, New York, NY, 1993, 2006.
4. Tung, P. and M. Mikasinovic. Eliminating Cavitation from Pressure-Reducing Orifices. *Chemical Engineering*, December 12, 1983.
5. TVA Mechanical Design Guide DG-M3.5.3. *Analysis and Control of Water Hammer in Large-Diameter Raw Water Piping Systems*. Tennessee Valley Authority, Knoxville, TN, 1982.

6. Streeter, V. and E. B. Wylie. *Hydraulic Transients*, McGraw-Hill, Inc., Newe York, NY, 1967.
7. Linsley, R. K. et al. *Water-Resources Engineering*, 4th ed., p. 367, McGraw-Hill, Inc., 1992.
8. Paramakian, J., *Water Hammer Analysis*, Prentice-Hall, Inc, New York, NY, 1955.
9. TVA Mechanical Standard Drawing SD-M6.1.1, *Typical System Air Release Requirements*, Tennessee Valley Authority, Knoxville, TN, 1981.

9 Heat Exchangers

9.1 SHELL-AND-TUBE HEAT EXCHANGERS

The detailed design of the HX is normally left to the HX manufacturer. However, the engineer is inevitably faced with ensuring that the HX will meet the needs of the plant over a wide range of operating conditions. Therefore, it is incumbent on the engineer once the HX is procured to be able to predict the performance of the HX over the range of anticipated conditions.

There are many types of HXs in power plants and industrial plants such as condensers, feedwater heaters, etc. The most common type of HX served by the cooling water system is the shell-and-tube heat exchanger (STHX) shown in Figure 9.1 consisting of tubes rolled or welded into tube sheets on both ends of the tubes. Cooling water enters the tubes through the inlet header, passes through the tubes where it increases in temperature, and passes out of the HX through the outlet header. The fluid to be cooled enters the space on the outside of the tubes (referred to as the shell) where it is directed by baffles to passes over the tubes several times to increase the heat transfer before exiting the shell.

There are many different configurations of STHX. The one illustrated in Figure 9.1 is a simple counter-flow HX in which both the cooling water and the fluid to be cooled pass through the HX only once. Other configurations include three and four passes on the tube side and two passes on the shell side with a divider plate in the middle of the shell.

From the first law of thermodynamics,

$$m_h \; h_{h-in} + m_c \; h_{c-in} = m_h \; h_{h-out} + m_c \; h_{c-out} \tag{9.1}$$

$$Q = m_h \left(h_{h-in} - h_{h-out} \right) = m_c \left(h_{c-out} - h_{c-in} \right) \tag{9.2}$$

where
 h_{c-in} = cold stream enthalpy in
 h_{c-out} = cold stream enthalpy out
 h_{h-in} = hot stream enthalpy in
 h_{h-out} = hot stream enthalpy out
 m_c = mass flow rate of cold stream
 m_h = mass flow rate of hot stream
 Q = rate of heat transfer.

Since

$$\Delta h = c_p \, \Delta T \tag{9.3}$$

127

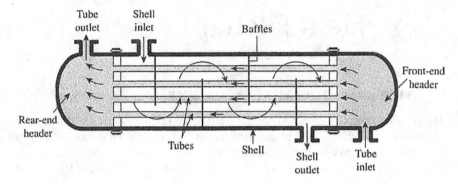

FIGURE 9.1   Shell-and-tube heat exchanger. (Courtesy of Thomas).

then

$$Q = m_h \, c_{p,h} \left(T_{h-in} - T_{h-out}\right) = m_c \, c_{p,c} \left(t_{c-out} - t_{c-in}\right) \qquad (9.4)$$

where

t_{c-in} = cold stream temperature in
t_{c-out} = cold stream temperature out
T_{h-in} = hot stream temperature in
T_{h-out} = hot stream temperature out
$c_{p,h}$ = hot stream specific heat
$c_{p,c}$ = cold stream specific heat.

Any one of the six parameters may be calculated if the other five are known. For example,

$$m_h = \frac{m_c \, c_{p-c} \left(t_{c-out} - t_{c-in}\right)}{c_{p-h} \left(T_{h-in} - T_{h-out}\right)} \qquad (9.5)$$

A commonly used method of HX analysis to determine the rate of heat transfer through a HX is the log-mean temperature difference method defined in Equation (9.6).

$$Q = U \, A_h \, F \, LMTD \qquad (9.6)$$

where

U = overall heat transfer coefficient
A_h = effective hot-side HX surface area
F = LMTD correction factor
$LMTD$ = log mean temperature difference

where

$$LMTD = \frac{\Delta t_1 - \Delta t_2}{Ln\left(\dfrac{\Delta t_1}{\Delta t_2}\right)} \qquad (9.7)$$

and

$$\Delta t_1 = T_{h,i} - t_{c,o} \qquad (9.8)$$

$$\Delta t_2 = T_{h,o} - t_{c,i} \qquad (9.9)$$

The accepted practice is to take the hot-side (shell-side) effective surface area, A_h, as the HX reference surface area where

$$A_h = N_{tubes}\, N_{pass}\, \pi\, d_o\, l_{eff} \qquad (9.10)$$

where
N_{tubes} = number of tubes per pass
N_{pass} = number of passes
d_o = outside tube diameter
l_{eff} = effective tube length

The LMTD correction factor, F, is a function of the type of HX. The value of "F" equal to one applies for counter-flow arrangements such as seen in Figure 9.1 and in other counter-flow HX arrangements. (See Section 9.2 and 9.3.) A formula for the value of "F" is available for a few other flow arrangements in the open literature.[1]

Another method of HX analysis is the effectiveness method as defined with reference to the cold stream as

$$Q = m_c\, c_{p-c}\left(t_{c-out} - t_{c-in}\right) = m_c\, c_{p-c}\, P\left(T_{h-in} - t_{c-in}\right) \qquad (9.11)$$

$$P = \frac{t_{c-out} - t_{c-in}}{T_{h-in} - t_{c-in}} \qquad (9.12)$$

where effectiveness, P, as defined in the equation above is the cooling range divided by the cooling potential.

A formula for the value of "P" is available for other flow arrangements in the open literature.[2]

The overall heat transfer coefficient, U, may be expressed as the inverse of the sum of the resistances to heat transfer through the following:

- exterior convection layer, r_h
- exterior fouling, $r_{f,h}$
- tube wall, r_w

- interior fouling, $r_{f,c}$
- interior convection layer, r_c

$$\frac{1}{U} = r_h + r_{f,h} + \left(\frac{A_h}{A_w}\right) r_w + \left(\frac{A_h}{A_c}\right) r_{f,c} + \left(\frac{A_h}{A_c}\right) r_c \qquad (9.13)$$

where

$$r_h = \frac{1}{h_h} \quad r_c = \frac{1}{h_c} \qquad (9.14, 9.15)$$

so that

$$U = \frac{1}{\dfrac{1}{h_h} + r_{f,h} + \left(\dfrac{A_h}{A_w}\right) r_w + \left(\dfrac{A_h}{A_c}\right) r_{f,c} + \left(\dfrac{A_h}{A_c}\right) \dfrac{1}{h_c}} \qquad (9.16)$$

where
h_h = hot-side convection coefficient
h_c = cold-side convection coefficient.

Note that all fouling resistances are referenced to the hot side (i.e. the shell side) of the HX tubes. The effective area of the wall and the wall resistance are shown as follows:

$$A_w = \frac{A_o - A_i}{\ln\left(\dfrac{A_o}{A_i}\right)} \qquad (9.17)$$

$$r_w = \frac{d_o - d_i}{2\,k_t} \qquad (9.18)$$

where d_o and d_i are the tube outside and inside diameters and k_t is the tube material thermal conductivity.

The value of the cold-side convection coefficient, h_c, is a function of the Nusselt number, Nu, as follows:

$$h_c = Nu \left(\frac{k_c}{d_i}\right) \qquad (9.19)$$

where
k_c = thermal conductivity of water.

Several equations are available for the value of the Nusselt number.[1] The commonly used classic Colburn[3] equation is as follows:

$$Nu = 0.023\ Re_t^{0.8}\ Pr_t^{1/3} \qquad (9.20)$$

where
  Re_t = tube-side Reynolds number,

$$Re_t = \rho v_t d_i / \mu_t \tag{9.21}$$

and
  Pr_t = tube-side Prandtl number,

$$Pr_t = c_{p,t}\mu_t / k_t \tag{9.22}$$

where
  ρ = density
  μ_t = average tube viscosity
  v_t = tube velocity.

No such similar direct method for calculating the hot-side convection coefficient, h_h, exists, as it is a complicated function of the HX shell geometry such as the type and spacing of the baffles, the leakage through the tube support plates, etc. Manufacturers generally hold this data to be proprietary. If a manufacturer's HX data sheet is available, the most direct method for calculating h_h is as follows:

$$h_{h,\ design} = \cfrac{1}{\cfrac{1}{U_{design}} - r_{f,h} - \left(\cfrac{A_h}{A_w}\right)r_w - \left(\cfrac{A_h}{A_c}\right)r_{f,c} - \left(\cfrac{A_h}{A_c}\right)\cfrac{1}{h_c}} \tag{9.23}$$

where the value of "U_{design}" is available or may be calculated from the data sheet.

If the design flow rate, inlet and outlet temperatures, number and size of tubes, and the value of "F" are known, then "U_{design}" may be calculated from Equation (9.24).

$$U_{design} = \frac{Q}{A_h\ F\ LMTD} \tag{9.24}$$

For HXs in cooling water systems, the value "h_c" may vary considerably during operation, as the cooling water flow and temperature often change seasonally as the inlet water temperature changes. The value of "h_h" may vary slightly from the design value during operation, but the change is normally minor compared with the other resistances in Equation (9.13), and in particular the tube-side fouling term that often predominates. Therefore, it is often acceptable to assume that the value for "h_h" and thus "r_h" remains constant in many cooling water system applications. However, if that is not the case, consult Reference 2.

9.2 AIR-TO-WATER HEAT EXCHANGERS

An air-to-water heat exchanger (AWHX) transfers heat between warm air and cooling water through an array of finned tubes as shown in Figure 9.2. AWHX are widely used to cool spaces inside power plants and other industrial plants and for many other applications such as to control the impact of an accident in a nuclear plant. A familiar example of an AWHX is the radiator in an automobile, and like that application, it is typically attached to a fan which draws air through a series of serpentine coils. As such it is classified as a cross-flow HX in which the LMTD correction factor "F" is equal to 1 if there are 4 or more passes through the coils as shown in Figure 9.2.

The convection boundary resistance to heat transfer on the air side of an AWHX can be as much as 20 to 30 times that on the tube side where there is turbulent flow of water through the tubes. To compensate for the poor heat transfer on the air side, fins are provided to increase the surface area on the air side by as much as 20 or more times that of the water side. As a practical manufacturing consideration, these fins are normally rectangular in shape. Accordingly, due to the resistance to conduction heat transfer down the fin to the surface of the water, increasing the length of the fin reaches a point of diminishing returns.[2] The measure of this reduction in heat transfer is referred to as the fin efficiency, η_{fin}. Some AWHX employ helical fins that are brazed to the tubes while others employ plate fins that are typically analyzed as equivalent fins.

The overall heat transfer coefficient at design conditions, U_{design}, is calculated similar to Equation (9.16) for STHXs as shown in Equation (9.25).

$$U = \frac{1/\eta_{fin}}{\dfrac{1}{h_h} + r_{f,h} + \left(\dfrac{A_h}{A_w}\right) r_w + \left(\dfrac{A_h}{A_c}\right) r_{f,c} + \left(\dfrac{A_h}{A_c}\right)\dfrac{1}{h_c}} \qquad (9.25)$$

Note that the only difference between Equation (9.19) and (9.25) is that the numerator contains the term for fin efficiency. In most AWHX designs, the fin efficiency is very nearly 1.0. If a more accurate analysis is required, consult Reference 2.

FIGURE 9.2   Air-to-water heat exchanger. (Courtesy of ChillX Chillers).

The tube-side convection coefficient may be computed as in Equation (9.19).

As was the case with STHXs and as shown in Equation (9.26), the air-side convection heat transfer coefficient for the AWHX may be found from vendor data to eliminate the need for hot-side heat transfer correlations specific to the geometry and configuration of the HX.

$$h_{h,\,design} = \frac{1/\eta_{fin}}{\dfrac{1}{U_{design}} - r_{f,h} - r_w - \left(\dfrac{A_h}{A_c}\right) r_{f,c} - \left(\dfrac{A_h}{A_c}\right)\dfrac{1}{h_c}} \tag{9.26}$$

As with the STHX, it is often acceptable to assume that the value for "h_h" and thus "r_h" remains constant in many cooling water system applications.

9.3 PLATE HEAT EXCHANGERS

Figure 9.3 illustrates a typical plate heat exchanger (PHX) consisting of a frame and a pack of corrugated plates separated from each other with gaskets and clamped together between two end covers with bolts. The cooling water and the fluid to be cooled flow through the channels created between the plates. The fluids enter and exit through ports located in the four corners of the plates, and the gaskets seal the plates at their outer edges and around the ports except as required to achieve the desired flow between the plates. A variety of flow patterns is made possible by judicious design of the gaskets. There are many plate designs. A very common plate design is the "herringbone" in which the plates are stamped with a chevron pattern on either a 30 or 60° angle with the horizontal and alternate plates inverted so that the plates form crisscross passages with frequent points of contact between the plates. PHX are compact and are generally easily cleaned with a wire brush in a single day, and their surface area can be easily increased by simply adding more plates. When compared to STHX, PHX weigh less, are less expensive, occupy less floor space (no

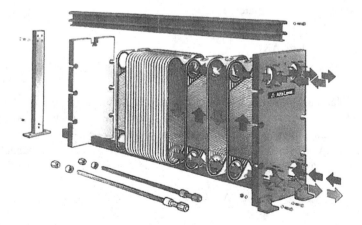

FIGURE 9.3 Plate heat exchanger. (Courtesy of Alfa Laval Corp.)

tube-pulling space required), and have higher coefficients of heat transfer. Therefore, they require less cooling water flow to achieve the same degree of cooling. PHX are primarily suited for liquid-to-liquid heat transfer, although working fluids involving single-phase gases are possible.

Alfa Laval, a PHX manufacturer, suggests the following performance limits in their literature:

Pressure, P (k Pa)	2500
Temperature, t (°C)	150
Total effective area, Aeff (m²)	2200
Mass flow rate, w (kg/sec)	1000

The analysis of PHX is similar to that of other HX. The overall heat transfer coefficient, U, may be expressed as the inverse of the sum of the resistances to heat transfer as shown in Equation (9.27).

$$U = \frac{1}{\frac{1}{h_c} + \frac{1}{h_h} + r_w + r_{fouling}} = \frac{Q}{A\ F\ (LMTD)} \qquad (9.27)$$

where

$r_{fouling}$ = is the sum of the fouling resistance on both sides of the plate.

Flow through adjacent passages in a PHX is countercurrent. Therefore, the LMTD correction factor, "F" is 1 for most PHX applications. The PHX consists of a single type of plate in which the geometry of the plate is the same on both sides.

The geometry of a PHX is defined by the following parameters:

Number of PHX in service, NHX
Height of plates, L_H
Width of plates, L_w
Compressed height of plates, L_{CP}
Number of plates, N_p
Thickness of plate, ΔX
Total effective area, A_{eff}.

Since the plates are corrugated, the effective area of a plate is greater than the product of the width times the height by the surface enlargement factor, ϕ, which can be as large as 1.5, depending on the plate manufacture and can easily be measured with a flexible tape measure. With this information, one may calculate the number of channel passes, the effective area, the spacing between plates, and the hydraulic diameter as follows:

$$N_{cp} = (N_p - 1)/2 \qquad (9.28)$$

$$A_{eff} = \phi(L_W\ x\ L_H) \qquad (9.29)$$

$$b = \left(\frac{L_c}{N_p}\right) - \Delta X \tag{9.30}$$

$$D_e = \frac{4 L_w b}{(2 L_w + 2b)} \tag{9.31}$$

The coefficient of heat transfer, h, is as shown in Equation (9.32).[4,5]

$$h_c = h_c = C \ \mathrm{Re}^{3/4} \ \mathrm{Pr}^{1/3} \left(\frac{k_c}{D_e}\right) \tag{9.32}$$

The Reynolds number is

$$\mathrm{Re} = D_e G / \mu \tag{9.33}$$

where μ is the viscosity and

$$G = \frac{m}{N_p \ b \ L_w} \tag{9.34}$$

and m is the mass flow rate.

The Prandtl number is

$$\mathrm{Pr} = \frac{\mu \ c_p}{k} \tag{9.35}$$

These terms are equally applicable to both the hot and cold sides of the PHX.

In 1999, Bowman[6] pointed out that since the geometry on both sides of the PHX is the same, the expression for "h" for both the hot and cold side is similar. Therefore, the value of "C" in Equation (9.26) is the same for both sides, and one may solve for "C" for a clean PHX as follows:

$$U = \frac{1}{h_c} + \frac{1}{h_h} + r_w = \frac{A \ (LMTD)}{Q} \tag{9.36}$$

$$\frac{1}{C \ \mathrm{Re}_c^{3/4} \ \mathrm{Pr}_c^{1/3} \left(\frac{k_c}{D_e}\right)} + \frac{1}{C \ \mathrm{Re}_h^{3/4} \ \mathrm{Pr}_h^{1/3} \left(\frac{k_h}{D_e}\right)} = \frac{A \ (LMTD)}{Q} - r_w \tag{9.37}$$

$$C \ \frac{\mathrm{Re}_c^{3/4} \ \mathrm{Pr}_c^{1/3} \ k_c}{D_e} + C \ \frac{\mathrm{Re}_h^{3/4} \ \mathrm{Pr}_h^{1/3} \ k_h}{D_e} = \frac{1}{\frac{A \ (LMTD)}{Q} - r_w} \tag{9.38}$$

$$C = \dfrac{\dfrac{D_e}{\mathrm{Re}_c^{3/4}\,\mathrm{Pr}_c^{1/3}\,k_c} + \dfrac{D_e}{\mathrm{Re}_h^{3/4}\,\mathrm{Pr}_h^{1/3}\,k_h}}{\dfrac{A\,(LMTD)}{Q} - r_w} \qquad (9.39)$$

This equation makes it possible to estimate the performance of a PHX at other than design conditions by simply knowing the physical parameters and the predicted heat transfer rate for a single set of design operating conditions (i.e. hot-side and cold-side mass flows and inlet and outlet temperatures). If the design data are not available, the value of C could be determined by conducting a series of heat transfer tests on a clean PHX over a range of operating conditions.

9.4   HEAT EXCHANGER FOULING

As discussed in Chapter 6, there are two forms of fouling in cooling water systems, microfouling and macrofouling. Macrofouling occurs in a HX when sediment or biological creatures such as Asiatic clams or zebra mussels or aquatic weeds such as seagrass block the flow through a HX. Macrofouling can have a significant impact on the ability of a HX to operate properly, as the cooling water flow blockage has the effect of reducing the effective surface area of the HX. (See Equation (9.6)) When some HX tubes are blocked, the flow through the remaining tubes tends to increase thus increasing the value of "U" in those tubes. The net effect is to reduce the overall heat transfer rate, Q, but not as much as the reduction in the effective area would indicate.

The importance of eliminating macrofouling from the cooling water system as discussed in Section 6.3 cannot be over-emphasized. In addition, provision should be made in the design to measure both the flow and pressure drop through the HX, so that macrofouling can be monitored in operation.

Although a HX is normally procured with a performance specification specifying the type of HX and the desired amount of cooling, the one additional parameter that must be specified by the engineer is the fouling resistance, r_f. The HX vendor cannot be expected to know the fouling conditions of the cooling water, $r_{f,c}$ and process side fluid, $r_{f,h}$. An estimate of "$r_{f,h}$" varies widely depending on the fluid that is being cooled and is beyond the scope of this book.

A distinction should be made between biological fouling and scaling on the tube side caused by a deposition of substances such as calcium or manganese dioxide that may plate out on the HX surface due to cooling water with a high pH and/or high temperature. The thickness of such a deposit may continue to increase without limit, whereas biological fouling may reach a point where the thickness of the fouling layer reaches an asymptotic limit due to the shearing effect of the cooling water on the surface. Somerscales[7] described asymptotic fouling (AF) as fouling where resistance to heat transfer initially increases rapidly when the tube is first exposed to cooling water but then decreases steadily until the fouling resistance is constant. AF is normally present when the fouling is due to biological microfouling but not due to sediment or scale buildup, which is hard and resistant to sloughing off.

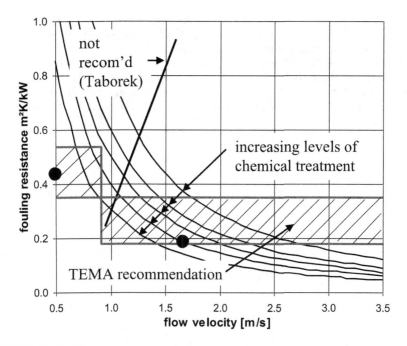

FIGURE 9.4   Fouling resistance.

For raw water, the Tubular Exchange Manufacturer's Association[10] (TEMA) recommends a minimum design fouling value of 0.002 to 0.003 hr-ft²-°F/Btu (0.00036 to 0.00054 s-m²-°C/J) for tube-side velocities below 3.0 ft/sec (0.91 m/s) and 0.001 to 0.002 hr-ft²-°F/Btu (0.00018 to 0.00036 s-m²-°C/J) for velocities above 3.0 ft/sec (0.91 m/s). However, Taborek[10] noted that the fouling rate is also a function of tube material and cooling water quality. Figure 9.4 shows the range of fouling rates recommended by Taborek, overlaid onto those recommended by TEMA for river water as well as data reported in References 8 and 9. (Taborek[10] stated that $r_f = f(V^{-1.75})$).

REFERENCES

1. ASME PTC 12.5-2000. *Single Phase Heat Exchangers*, September 2000.
2. Thomas, L. C. *Heat Transfer- Professional Version*, 2nd ed., Capstone Publishing Corp., Tulsa, OK, 1999.
3. Colburn, A. P. A. Method of Correlating Forced Convection Heat Transfer Data and a Comparison with Fluid Friction. *Transactions of AIChE*, Vol. 29, pp. 174–219.
4. Rauj, K. S. N. and J. Chand. Consider the Plate Heat Exchanger. *Chemical Engineering*, Vol. 87, No. 16, pp. 133–144, August 11, 1980.
5. Arpaci, Vedat S. Microscales of Turbulent Heat and Mass Transfer. In *Advances in Heat Transfer*, Academic Press, Cambridge, MA, 1997, pp. 1–91.
6. Bowman, C. F. *Plate Heat Exchangers. Electric Power Research Institute Service Water System Reliability Improvement Seminar*, Biloxi, MS, 1999.
7. Somerscales, E. F. C. Fouling of Heat Transfer Surfaces: An Historical Review. *Heat Transfer Engineering*, Vol. 11, No. 1, pp. 19–36, 1990.

8. Nolan, C. M. and B. H. Scott. *On Line Monitoring of Heat Exchangers Microfouling: An Alternative to Thermal Performance Testing. EPRI SW Reliability Improvement Seminar*, Electric Power Research Institute, Palo Alto, CA, 1995.
9. Zelver, N. et al. *Tube Material, Fluid Velocity, Surface Temperature and Fouling: A Field Study*. CTI Paper TP-84-16, Cooling Tower Institute, Houston, TX, 1984.
10. Taborek, J. *Assessment of Fouling Research on the Design of Heat Exchangers. Fouling Mitigation of Industrial Heat Exchangers Conference*, Shell Beach, CA, 1995.

10 Heat Rejection

10.1 REGULATORY REQUIREMENTS

In 1972, the United States Congress passed the CWA, establishing the basic structure for regulating discharges of pollutants including heat through pipes or man-made ditches into the waters of the United States. Under the CWA, the EPA was charged with implementing these pollution control measures and with developing national water quality criteria recommendations. The EPA has implemented the program through the NPDES. Each state was required to establish limits for thermal discharge in their state that are intended to protect the propagation of the receiving body of water's balanced indigenous population of fish, shellfish, and wildlife. These limits address the following parameters outside a reasonable mixing zone near the point of discharge from a plant:

- maximum discharge temperature
- maximum rate of change in discharge temperature
- maximum temperature rise by the plant.

Under Section 316(a) of the CWA, the permitting state may impose alternative effluent limitations for controlling thermal discharges in lieu of the effluent limits that would otherwise be required if the organization receiving the permit can demonstrate that the otherwise applicable thermal discharge effluent limit is more stringent than necessary to assure the protection and propagation of the body's balanced indigenous population of fish, shellfish, and wildlife. Under the CWA, "new source performance standards" set the level of allowable discharges from new industrial facilities. This provision requires the EPA to determine the best available demonstrated control technology for a given industry that may be more stringent than the limits for existing dischargers up to and including a "zero discharge" standard.

As stated in Section 3.3, Section 316(b) of the CWA requires power and industrial plants drawing water from the waters of the United States to implement the BTA to reduce injury or death of fish and other aquatic life and requires new plants at existing facilities to install closed-cycle recirculating systems or to reduce actual intake flow to a level commensurate with that attained by a closed-cycle recirculating system.

Therefore, future cooling water systems may require some method of rejecting heat directly to the environment such as cooling towers or a spray pond to either operate as a closed system or in a "helper" mode as discussed in Section 7.6.

10.2 COOLING TOWERS

10.2.1 CROSS-FLOW MECHANICAL DRAFT COOLING TOWERS

Figure 10.1 Illustrates a common type of cross-flow mechanical draft cooling tower (MDCT).

139

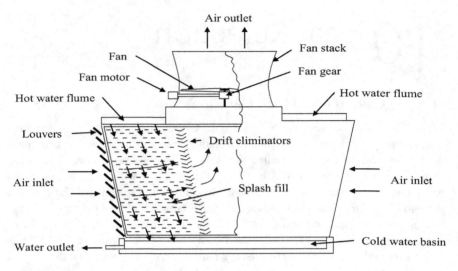

FIGURE 10.1 Cross-flow mechanical draft cooling tower.

A cross-flow MDCT consists of one or more cells each of which consists of a large fan affixed on the top of a structure containing splash-type fill material. Warm cooling water is pumped up to the top of the structure and distributed in the hot water flume atop the structure. It then passes through specially designed nozzles embedded in the floor of the hot water flume. The nozzles are fitted with splash plates to evenly distribute the cooling water over the splash-type fill below. The motor-driven fan atop each cell of the MDCT operates to draw air into the side of the structure through louvers. The air passes across the water as it is continually separated into fine drops by the splash fill before falling to the cold water basin below. After interacting with and cooling the falling cooling water droplets, the air passes through drift eliminators and out through the fan at the top of the MDCT. The drift eliminators are designed to minimize the number of fine drops of cooling water carried out of the fill section of the MDCT and passed through the fan. A fan stack surrounds each fan to increase the efficiency of the fan by minimizing recirculation of the air. The cold cooling water is collected from all of the MDCT cells in a single cold-water basin located below all of the MDCT cells and then returned to the IPS.

10.2.2 CROSS-FLOW NATURAL DRAFT COOLING TOWER

Figure 10.2 Illustrates a cross-flow natural draft cooling tower (NDCT). Typically a cooling water system alone would not be sufficiently large to require a NDCT, but since in the past they have been employed in conjunction with a power plant CCW system and since in power plants the cooling water is frequently discharged into the CCW system, the NDCT is discussed here for completeness.

A cross-flow NDCT operates in a manner similar to a cross-flow MDCT except that the air flow is created by the chimney effect of a tall veil. Even when there is no CCW flow to the NDCT, air flow is induced by the lower density of the air at the top

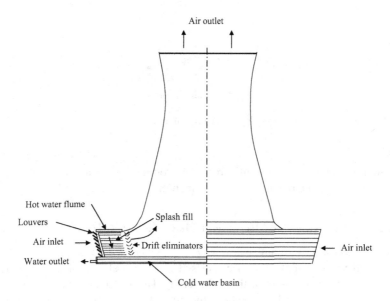

FIGURE 10.2 Cross-flow natural draft cooling tower.

of the veil. When the power plant is not operating but there is still CCW flow to the NDCT, there is less dense saturated air entering the veil, since the molecular weight of water vapor (18) is less than that of air (29). Depending on the water loading on the fill section, this may result in an increase in air flow as the lighter air tends to rise. When the power plant is operating, the heat added to the CCW results in an even greater air flow, as warm saturated air is even less dense. A canopy between the donut-shaped heat transfer section around the veil containing the splash-type fill material and the veil encloses the interior of the NDCT so that no air can bypass the fill section.

10.2.3 SPLASH-TYPE COOLING TOWER FILL MATERIAL

Figure 10.3 illustrates splash-type fill that is commonly used with cross-flow MDCT and NDCT.

Splash-type fill material may be constructed from a variety of materials such as wood slats but most commonly of plastic in large cross-flow cooling towers serving large plants. The plastic slats may be oriented either parallel to or perpendicular to the air flow. (The parallel orientation results in less resistance to air flow.) The slats, normally supported by a system of plastic or coated wire hangers, are frequently perforated and come in a variety of shapes for strength.

One of the major drawbacks with cross-flow cooling towers, both MDCT and NDCT, is the propensity for ice to build up on the fill material in cold weather. Various schemes have been employed to prevent ice buildup such as diverting a larger portion of the CCW to the perimeter of a NDCT or shutting off or reversing the fan flow in MDCT, but all such schemes require diligent observation to identify when

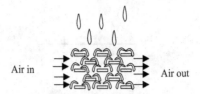

FIGURE 10.3   Splash-type cooling tower fill material.

ice buildup has begun. As a result, cross-flow cooling towers are in somewhat disfavor where there is the potential for ice formation.

10.2.4   COUNTER-FLOW MECHANICAL DRAFT COOLING TOWER

Figure 10.4 Illustrates a counter-flow MDCT. As is the case with a cross-flow MDCT, each cell in a counter-flow MDCT consists of a large fan affixed on the top of a structure but the fill material is of the thin-film variety. The heated cooling water is pumped through a piping system located above the fill section and passes through especially designed nozzles located at the bottom of pipes fitted with splash plates to evenly distribute the cooling water over the thin film fill material below. As with a cross-flow MDCT, the motor-driven fan atop each cell operates to draw air into the side of the structure. The air passes through the fill material where the cooling water is running down the fill material in thin sheets before it falls to the cold water basin below. After interacting with and cooling the falling cooling water, the air passes through drift eliminators and out through the fan at the top of the MDCT. The cold cooling water is collected from all of the MDCT cells in a single cold water basin located below all of the MDCT cells and returned to the IPS.

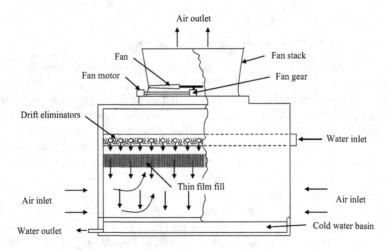

FIGURE 10.4   Counter-flow mechanical draft cooling tower.

10.2.5 COUNTER-FLOW NATURAL DRAFT COOLING TOWER

Figure 10.5 Illustrates a counter-flow NDCT.

In an NDCT, the heated CCW is pumped up through concrete standpipes that distribute the CCW through a piping system located above the fill section. It then passes through especially designed nozzles located at the bottom of the pipes fitted with splash plates to evenly distribute the CCW over the thin film fill material below. A counter-flow NDCT operates in a manner similar to a counter-flow MDCT except that as in the case of a counter-flow NDCT, the air flow is created by the chimney effect of a tall veil. As with a cross-flow NDCT, even when there is no CCW flow to the NDCT air flow is induced by the difference in the density of the air at the top of the veil, which is less than that at ground level. Air flow is also increased when the power plant is operating since the heat added to the CCW results in warm saturated air that is even less dense. Counter-flow NDCT have proven to be much more resistant to ice damage due to the much more durable fill material and the fact that damage is normally limited to the outer edges of the HX section.

10.2.6 THIN FILM-TYPE COOLING TOWER FILL MATERIAL

Figure 10.6 illustrates thin film-type fill commonly used with counter-flow MDCT and NDCT.

One type of thin film-type fill shown in Figure 10.6 consists of several layers of very thin vertical sheets hung from structural members such that when the cooling water lands on the top of the top sheet, it tends to dribble down the sheet in a thin film

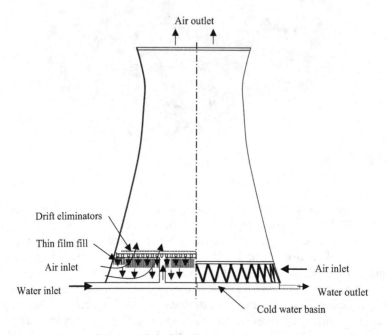

FIGURE 10.5 Counter-flow natural draft cooling tower.

Air out

FIGURE 10.6   Thin film-type cooling tower fill material.

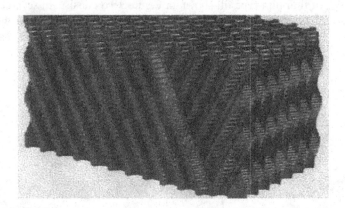

FIGURE 10.7   Plastic cooling tower fill material.

before falling to the basin below. Another type of cooling tower fill material shown in Figure 10.7 is thin corrugated plastic glued together in a herringbone pattern described as "egg crates" manufactured by Munters and Hitachi. This type of fill material, though more efficient than that shown in Figure 10.6, may be susceptible to clogging. If the surface of the plastic is permitted to be covered with pseudomonas bacteria (i.e. slime), suspended solids in the cooling water tend to build up on the surface of the plastic, eventually obstructing flow. Chemical treatment systems employing oxidizing biocides such as chlorine are volatile, and the biocide and may not control the buildup of pseudomonas on the fill material.

10.3 SPRAY PONDS

10.3.1 ADVANTAGES OF SPRAY PONDS

A spray pond is a system of pipes and spray nozzles that spray water into the air to cool the water. They are similar in this way to cooling towers in that they dissipate waste heat to the atmosphere principally through evaporation. Much of the advance in the understanding of the design of spray ponds has been gained because they play a critical role in the safe operation of some nuclear plants.[1] Unlike MDCT, spray ponds need no fan power or fill maintenance. Spray ponds are less susceptible to ice damage than cooling towers and can operate satisfactorily under cold conditions if provided with a spray bypass to maintain minimum temperatures in the pond.

10.3.2 CONVENTIONAL FLAT BED SPRAY PONDS

Figure 10.8 shows the conventional flat bed spray pond (FBSP) that was the ultimate heat sink for the now defunct Rancho Seco nuclear station.[1] Typically, a FBSP consists of a series of trees mounted on straight header pipes in a rectangular pattern with each tree consisting of a riser pipe and four cross arms at 90 degree angles approximately five feet (1.5 m) long with a spray nozzle at the end of each cross arm pointed vertically. The dimensions of each of the two FBSP at Rancho Seco is 165 by 330 feet (50.3 by 100.6 m), with 304 nozzles each delivering 53 gal/min (3.34 l/s) at 7 Lbf/in^2 (48.3 kPa) nozzle pressure.

In the FBSP design with all spray nozzles oriented in the vertical direction, the bulk drag force of the water droplets (vertically downward) resists the natural

FIGURE 10.8 Spray nozzles in a conventional flat bed spray pond.

buoyancy of the warm air (which is of the same order of magnitude and is directed vertically upward). Since the two forces acting on the air oppose each other, the result is a reduced air flow rate through the spray region and a large increase in the local wet-bulb temperature (WBT). Additionally with a large FBSP, the falling water tends to block air flow into the central parts of the array, requiring an interference factor to be applied to determine the local WBT as a function of the ambient WBT.[1]

10.3.3 ORIENTED SPRAY COOLING SYSTEM

The oriented spray cooling system (OSCS) employs a radically different spray pond design that overcomes these problems. Figure 10.9 shows the OSCS that was constructed, operated, and thoroughly tested at the Columbia Generating Station (CGS).

The outstanding feature of the OSCS design is the circular arrangement of the spray nozzles on spray trees. The spray trees are spaced approximately 13'-9" (4.2 m) on center with Spraying Systems Co. Whirljet Type $1\frac{1}{2}$ CX SS 27 spray nozzles arranged in a helical pattern spaced approximately 4 feet (1.2 m) from the mast of the tree and approximately 2'-8" (0.81 m) apart vertically and oriented at a 35° angle from the vertical toward the center of the circle.[1] In this design, both the bulk drag force of the water droplets on the air and the buoyant force promote ventilation of the spray region, and ambient air flow to the spray nozzles is not obstructed. The result is a reduction on the local WBT in the spray region and improved cooling of the droplets as they fall through the spray region to the pond surface below. However, the required nozzle pressure is higher than with a FBSP.

FIGURE 10.9 Oriented spray cooling system at the Columbia Generating Station.

10.3.4 ANALYSIS OF SPRAY PONDS

The simplest measure of thermal performance of a spray pond is the efficiency, η, of the device as defined by the following standard formula:

$$\eta = \frac{t_1 - t_2}{t_1 - t_{WB}} \tag{10.1}$$

where

η = spray efficiency
t_1 = temperature of the water entering the spray nozzle
t_2 = temperature of the sprayed water
t_{WB} = ambient WBT.

Assuming a constant heat load and WBT one may solve for t_2 as follows:

$$t_2 = t_1 - \eta\left(t_1 - t_{WB-local}\right) \tag{10.2}$$

where

$t_{WB-local}$ = wet-bulb temperature at the spray nozzle.

10.3.5 SPRAY POND TEST RESULTS

Only a few quality full-scale spray pond efficiency tests are available in the open literature.[1] Figures 10.10 and 10.11 show the comparison among tests conducted on the conventional FBSP at Rancho-Seco and Okeelanta Florida complex and the

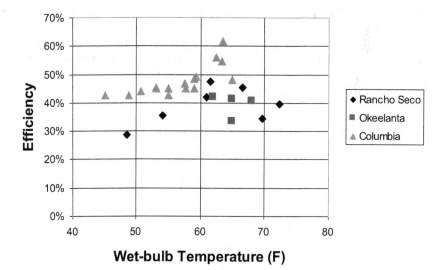

FIGURE 10.10 Flat bed spray ponds and oriented spray cooling system efficiencies vs. WBT.

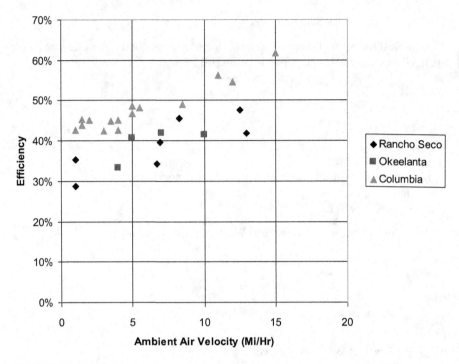

FIGURE 10.11 Flat bed spray ponds and oriented spray cooling system efficiencies vs. ambient air velocity.

OSCS at the CGS. One may see from Figure 10.10 that the spray efficiency increases with increasing WBT. This is the case for any type of evaporative cooling device such as a cooling tower.

Figure 10.11 shows the significant impact that ambient air velocity has on spray ponds, as the spray efficiency increases with increased ambient air velocity. At zero air velocity, the OSCS is seen to have a very significant advantage in efficiency, because the OSCS creates its own wind.

REFERENCE

1. Bowman, C. F. and S. N. Bowman *Thermal Engineering of Nuclear Power Stations: Balance of plant Systems*, CRC Press, Boca Raton, FL, 2021.

Nomenclature

Symbol	Definition	English Units	SI Units
A	corrosion allowance	in	m
A	wave speed	ft/sec	m/s
a_1	constant	Dimensionless	Dimensionless
a_2	constant	Dimensionless	Dimensionless
A	cross-sectional area	ft^2	m^2
A_{eff}	effective surface area	ft^2	m^2
A_h	effective hot-side heat exchanger surface area	ft^2	m^2
BHP	brake horsepower	Hp	Hp
C	saturation factor	Dimensionless	Dimensionless
C	Hazen-Williams "C" factor	Dimensionless	Dimensionless
c_p	specific heat	Btu/lbm-°F	kJ/kg-°C
$c_{p,c}$	tube-side specific heat	Btu/lbm-°F	kJ/kg-°C
$c_{p,h}$	hot-side specific heat	Btu/lbm-°F	kJ/kg-°C
D	pump bell diameter	ft	m
$d*$	dimensionless parameter	Dimensionless	Dimensionless
$d_{NOM, n}$	nominal pipe diameter in each section of pipe	in	m
d_{NOM}	nominal inside diameter of new pipe	in	m
d_{CALC}	calculated inside diameter of pipe	in	m
Δd_{MEAS}	measured diameter reduction	in	m
D	pipe diameter	ft	m
D_e	hydraulic diameter	ft	m
D_o	outside diameter	min	
d	diameter of pipe	in	
d_i	inside diameter of a pipe or tube	ft	m
d_o	tube outside diameter	ft	m
d_1	orifice diameter	in	
E_M	bulk modulus of elasticity of the mixture of water and air	lbf/in^2	Pa
Ep	bulk modulus of elasticity of the pipe	lbf/in^2	Pa
E_w	bulk modulus of elasticity of water	lbf/in^2	Pa
e	absolute roughness of pipe	in	
f	friction factor.	Dimensionless	Dimensionless
f	magnitude of the reflected pressure wave	lbf/in^2	kPa
F	LMTD correction factor	Dimensionless	Dimensionless
F	magnitude of the pressure wave	lbf/in^2	kPa
F_d	Froude number	Dimensionless	Dimensionless
g	gravitational constant	ft/sec^2	m/s^2
G	flow rate	gal/min	l/s
h_L	total head loss	ft	m
$h_{L,f}$	friction head loss	ft	m
h_c	cold-side convection coefficient	Btu/hr-ft^2-°F	J/s-m^2-°C
$h_{c\text{-}in}$	cold stream enthalpy in	Btu/lbm	kJ/kg
$h_{c\text{-}out}$	cold stream enthalpy out	Btu/lbm	kJ/kg
h_h	hot-side convection coefficient	Btu/hr-ft^2-°F	J/s-m^2-°C
$h_{h,design}$	design hot-side convection coefficient	Btu/hr-ft^2-°F	J/s-m^2-°C
$h_{h\text{-}in}$	hot stream enthalpy in	Btu/lbm	kJ/kg
$h_{h\text{-}out}$	hot stream enthalpy out	Btu/lbm	kJ/kg
H	pressure head	ft	kPa

(Continued)

Symbol	Definition	English Units	SI Units
H	total head	ft	m
$h_{L,F}$	form head loss per equivalent foot of pipe	ft	m
$h_{L,f/100}$	friction head loss in feet per 100 ft of pipe	ft	m
H_o	initial pressure head	ft	kPa
K_{air}	Henry's Law coefficient	Dimensionless	Dimensionless
K'_{air}	Henry's Law coefficient based on the HX outlet temperature	Dimensionless	Dimensionless
K_i	cavitation index	Dimensionless	Dimensionless
K_x	Henry's Law coefficient for gas x	Dimensionless	Dimensionless
k	resistance coefficient	Dimensionless	Dimensionless
k	thermal conductivity of water	Btu/hr-ft-$^\circ$F	kJ/s-m-$^\circ$C
k_c	thermal conductivity of water	Btu/hr-ft-$^\circ$F	kJ/s-m-$^\circ$C
k_t	thermal conductivity of the tube material	Btu/hr-ft-$^\circ$F	kJ/s-m-$^\circ$C
L	length of pipe	ft	m
L_{CP}	compressed height of plate	in	m
l_{eff}	effective tube length	ft	m
L_H	height of plate	in	m
$LMTD$	log mean temperature difference	$^\circ$F	$^\circ$C
L_w	width of plate	in	m
m	mass	lbm	kg
m	cooling water mass flow rate	lbm/hr	kg/s
MHP	Motor horsepower	hp	Hp
m	mass flow rate	lbm/hr	kg/s
m_c	mass flow rate of cold stream	lbm/hr	kg/s
m_h	mass flow rate of hot stream	lbm/hr	kg/s
N	rotational speed	rev/min	
N_{air}	mole fraction of all dissolved gases	Dimensionless	Dimensionless
$N'air =$	mole fraction that local conditions will permit to remain in solution	Dimensionless	Dimensionless
$N_{air\text{-}max}$	maximum mole fraction of air in the entering cooling water	Dimensionless	Dimensionless
N_x	mole fraction of dissolved gas, x	Dimensionless	Dimensionless
n	specific speed	Dimensionless	Dimensionless
n	number of moles	Dimensionless	Dimensionless
NHX	number of PHX in service	Dimensionless	Dimensionless
N_p	number of plates	Dimensionless	Dimensionless
N_{pass}	number of passes	Dimensionless	Dimensionless
$NPSHA$	net positive suction head available	ft	m
$NPSH_i$	cavitation inception	ft	m
$NPSHR$	net positive suction head required	ft	m
N_{tubes}	number of tubes per pass	Dimensionless	Dimensionless
Nu	Nusselt number	Dimensionless	Dimensionless
P	effectiveness	Dimensionless	Dimensionless
P	pressure	lbf/in^2	kPa
$P_{required}$	minimum gauge pressure to keep air in solution	lbf/in^2	kPa
P'	local static pressure	lbf/in^2	kPa
P_1	gauge pressure 1	lbf/in^2	kPa
P_2	gauge pressure 2	lbf/in^2	kPa
P	internal pipe pressure	lbf/in^2	kPa
P_b	barometric pressure	lbf/in^{2}A	kPaa
P_d	local downstream pressure	lbf/in^2	kPa

(Continued)

Symbol	Definition	English Units	SI Units
$P_{sat} =$	saturation vapor pressure at the maximum operating temperature	lbf/in^{2}A	kPaa
P_u	local upstream pressure	lbf/in^2	kPa
P_v	vapor pressure at the cooling water temperature	ft	M
P_x	partial pressure of dissolved gas x	lbf/in^{2}A	kPaa
Pr	Prandtl number	Dimensionless	Dimensionless
Q	Flow rate	gal/min	
Q_n	flow in each section of pipe	gal/min	
Q	maximum HX heat load	Btu/hr	kJ/s
Q	rate of heat transfer	Btu/hr	kJ/s
R	Universal gas constant	ft-lbf/lbm-mole °R	cal/gm-mole °K
Re	Reynolds number	Dimensionless	Dimensionless
Re_t	tube-side Reynolds number	Dimensionless	Dimensionless
$r_{fouling}$	fouling resistance on both sides of the plate	hr-ft^2/Btu	m^2-°C/W
S	maximum allowable stress	lbf/in^2	
S_{min}	minimum submergence	ft	m
s	suction specific speed	Dimensionless	Dimensionless
s	number of pipe sections	Dimensionless	Dimensionless
S	submergence	ft	m
T	absolute temperature	°R	°K
T'	absolute local temperature	°R	°K
T	time	sec	s
t	thickness of pipe wall	in	m
t	temperature	°F	°C
t_M	minimum required wall	in	
T_{out}	HX outlet temperature	°F	°C
T_{in}	HX inlet temperature	°F	°C
T_1	temperature of water entering the spray nozzle	°F	°C
T_2	temperature of sprayed water	°F	°C
t_{c-in}	cold stream temperature in	°F	°C
t_{c-out}	cold stream temperature out	°F	°C
T_{h-in}	hot stream temperature in	°F	°C
T_{h-out}	hot stream temperature out	°F	°C
t_{WB}	ambient wet-bulb temperature	°F	°C
$t_{WB-local}$	wet-bulb temperature at the spray nozzle	°F	°C
U	overall heat transfer coefficient	Btu/hr-ft^2-°F	W/m^2-°C
U_{design}	design overall heat transfer coefficient	Btu/hr-ft^2-°F	W/m^2-°C
V	volume	ft^3	m^3
V_{pipe}	water velocity in a pipe or tube	ft/sec	m/s
V_r	air-to-water volume ratio	Dimensionless	Dimensionless
V_S	velocity at suction inlet	ft/sec	m/s
VH_1	velocity head 1	ft	m
VH_2	velocity head 2	ft	m
VM_w	volume of water	ft^3	m^3
VM_a	volume of air	ft^3	m^3
VM_T	total volume	ft^3	m^3
V	velocity	ft/sec	m/s
V_o	initial fluid velocity	ft/sec	m/s
v_t	tube velocity	ft/sec	m/s
w	mass flow rate	lbm/sec	
y	temperature coefficient	Dimensionless	Dimensionless
Z_1	elevation of the gauge 1	ft	m

(*Continued*)

Symbol	Definition	English Units	SI Units
Z_2	elevation of the gauge 2	ft	m
ΔP	pressure drop across the orifice	lbf/in^2	kPa
ΔX	thickness of plate	ft	m
η_m	motor efficiency	Dimensionless	Dimensionless
η	spray efficiency	Dimensionless	Dimensionless
η_{fin}	fin efficiency	Dimensionless	Dimensionless
η_p	pump efficiency	Dimensionless	Dimensionless
μ	dynamic viscosity of the water	lbm/ft-hr	kg/m-s
μ_t	tube-side viscosity	lbm/ft-hr	kg/m-s
Δ	density	lbm/ft^3	kg/m^3
ρ_a	density of air	lbm/ft^3	kg/m^3
ρ_w	density of water	lbm/ft^3	kg/m^3
ρ_M	density of the mixture of water and air	lbm/ft^3	kg/m^3
σ	Thoma's cavitation constant	Dimensionless	Dimensionless

Index

Page numbers in **bold** indicate tables, page numbers in *italic* indicate figures.

Printed in the United States
by Baker & Taylor Publisher Services